MW00509511

THROUGH
THE
LEAVES

MARCH, 1919

Published Monthly By
THE GREAT WESTERN SUGAR CO.

DAILY CALL PRESS
LONGMONT, COLORADO
1919

Editors' Page

 THE opening of spring in any farming community means the beginning of new opportunities. In looking back over the work of the year previous, we can all see where we might have done a little better and as the new crop year opens up, we are all glad to have a chance once again to see if this time we cannot secure the desired results.

The more we study reasons for successful crops, the more we see the value of attention to the small details.

The man who has taken advantage of winter days to put his machinery in good shape so his spring work goes along without delay has gotten some of the most annoying farm troubles out of the way. The man who starts his work early and keeps ahead during the season is the man who is in position to take advantage of favorable climatic conditions which mean much in making satisfactory yields.

In this connection we want to call attention to the article by Mr. John Comer in this issue. Mr. Comer is superintendent of our beet seed department and is one of the best posted and most practical men we know of when it comes to general farming practice.

— — —

It is with considerable pride that we are able to present as our pictorial section this time, the pictures of all the sixteen factories owned and operated by the Great Western Sugar Company. The writer can look back over eighteen years and see the first factory at Loveland in 1901 and as the steady growth has continued during the intervening years, the confidence of the farmers in the integrity of the Company has grown with it.

What I Would do if I Were Going to Plant Sugar Beets

John Comer

F my land was plowed last fall, it will not usually be necessary to disk it. As a rule the use of a harrow and float will put it in good shape, especially if there has been a reasonable amount of moisture during the winter as there has been during the past one. If, however, the ground has had a good, heavy coat of manure or stubble has been plowed under, then I would use a disc in order to close up any air spaces which may be caused by the manure or stubble. I would work my soil after each rain during the spring to conserve the moisture. If I do this I find I have no trouble in making a good seed bed, which usually means a good stand of beets.

If my ground comes through the winter in a cloddy condition, I would roll it or level it, thus crushing the clods into the surrounding soil where I leave them until I get more moisture which will soften them so they will pulverize when harrowed. If these clods are left on top of the ground, it is a good deal like rain on a tin roof; it runs off without penetrating and instead of one rain softening them, so they will work, it will take several.

Breaking Alfalfa For Beets in the Spring

If I have alfalfa ground that I want to put into beets this spring, I would crown it not over two inches deep if possible. This can be done by keeping the plow lays sharp and not trying to cut too wide a furrow. When alfalfa ground is going into beets this spring, this crowning should be the first operation on the farm.

I can usually crown alfalfa in the spring two to three weeks before I can do the plowing, as the frost goes out to a depth of two or three inches long before other plowing can be done. If I get my alfalfa crowned in the month of March, which I usually can do, I follow immediately after with a spring tooth or other harrow, in order to get the crown out on the surface of the ground, and then I aim to have it replowed not later than April 10th, so as to get my beets planted by April 20th. The depth of this second plowing will depend on the depth to which the ground had been plowed before being seeded down, but it is not generally advisable to go more than seven to eight inches.

General Spring Plowing

Any other land that I have for spring plowing is always disked before plowing. This disking can also usually be done early before the ground is ready to plow. Spring plowing should always be done as early as possible in order to get the spring rains on it before the planting season and in order that the ground may settle and become firm. The earlier the plowing, the nearer will the results be to fall plowing.

All ground that is spring plowed, I like to have harrowed at least three times the same day it is plowed and the following day when it has "dusted off" on top so the float will not clog, it should be floated. Any clods that have been left by the harrow will be mashed into the soil by the float so the first moisture that comes will melt them.

If this does not pack the ground enough to make my seed bed sufficiently firm for planting, then I would roll and harrow before planting. In order to secure a good stand it is necessary that the seed bed be firmly packed.

The above is for early spring plowing. If I have any plowing to do late in the spring, say the middle or last of April, then I try to make a good seed bed with the moisture I have when the land is plowed. This usually means that I must harrow, float, roll and reharrow possibly within two days after the ground is plowed. It is quite essential that all beets be planted during the month of April if it is at all possible to do so, as early beets will make a better tonnage.

If your ground is sandy and liable to blow, it will have to be handled a little differently than I have outlined. This kind of soil should be plowed in the spring if possible when it is quite wet. On this kind of ground it is usually best to work to have some clods, but they should be small ones. After sandy ground is plowed quite wet it should lay perhaps two or three days before any further work is done, in order that small clods may form to keep it from blowing.

I usually find that the greatest trouble with sandy ground is that it cannot be worked enough to pack it sufficiently firm to make a good seed bed, without making it too fine so that it blows away. A good even job of plowing on this kind of land usually makes it unnecessary to use the float. Floating sandy ground usually causes it to blow, but after being plowed wet, if you will use a Western Land Roller, say three days after plowing, you will have a good many small clods. Immediately after it has been rolled I would harrow it with the teeth set straight up to break every little "sole" that the roller may have formed two or three inches under the surface. If your seed bed is still not firm enough this operation of rolling and harrowing should be repeated.

Further Seed Bed Preparation

All ground to be planted to beets should be harrowed before planting with the harrow teeth set slanting at about 45 degrees; the harrowing to be done in the opposite direction to that which you plant so the drill rows will cross the harrow marks at right angles. I would not plant after a rain without harrowing the ground.

131

Planting Beets

When I get ready to plant I always make sure that my drill is in good shape before taking it to the field. Never use a drill with worn out shoes. I find that if I have been using a drill with worn out shoes and then put on new shoes that I have to watch it very closely or I will get my seed in too deep. I always try to get my seed planted not more than one inch below the packed surface of the ground under the press wheel. If I do not have moisture enough to bring up the seed at that depth of planting I do not try to go down to moisture but plant it at this depth and wait for a rain, as beets that are planted deep will not come through any crust whatever and it always takes them so long to come up that they do not grow as they should for some time. In turning on the end rows I find it necessary to be very careful in backing up the drill to see that I do not fill the bottom of the shoes with dirt and thus clog the drill so it will not plant. I also watch my press wheels to see that they are tight and set at the right place to hold the shoes at the right depth.. I have often seen drills in fields with part of the press wheels at the top of the slot and others at the bottom. You can readily see that this would give an uneven planting, and when the beets come up sometimes two rows are good and two are poor. This is generally due to the way the press wheels are set. A good many farmers are liable to figure that by setting their lever one notch deeper they will plant their beets deeper, but this has nothing to do with the depth whatever. The depth is all gauged by the setting of the press wheels. The deeper the lever is set, the more pressure on the press wheels.

If I find when I have gotten to the end of the field that I have missed one row, I turn around and plant the whole four rows over again. I always make sure that my marker is right so that the guess row is as near 20 inches as possible. I find that if the rows are planted even, it helps a great deal when it comes to cultivating. I also find that if I have set my drill for one field that when I finish and go to another one that it is usually necessary to reset the press wheels to get the seed planted to the proper depth. I always get off my drill when I start planting in a new field and examine carefully the depth to which the seed is going. There are no two fields where the ground is packed exactly alike and it is almost impossible to set a drill to plant several fields without changing, as the depth of planting has a great deal to do with whether the wheels that carry the drill sink into the soil one inch or two inches, or whatever it is packed solid enough for them to set on top. You can readily see that if the wheels that carry the drill sink in on soft ground an inch or two deeper than they did in the field that is firmly packed that the shoe will doubtless be running considerably deeper. As the depth of planting is one of the most essential things to the securing of a good stand, it pays to watch it very closely.

132

News from Factory and Field

Longmont Factory

More than 20,000 acres of beets have been signed for this factory at this time and the end is not in sight. We hope all will put forth their greatest effort to make the year a record one for tonnage as well as acreage.

* * *

We have been called into many new districts this year to establish receiving stations where beets have not been grown before. We asked a man in one of these new districts why they were anxious to have the beet industry in their community. His reply was. "The reason I want to grow sugar beets is because those fellows that raise beets have more money than I do." He said further: "I went to a dairy sale the other day to buy some cows, and the beet farmers had so much money they just naturally bid so high for the stuff that I had to go back and sit down."

That is the story of the sugar beet in a nutshell, and it is only necessary to go into districts that have not been growing beets to learn the value of the crop.

* * *

The cattle feeders in this district have almost without exception had a very prosperous year, and are pleased with their results. Our pulp allotments, we believe, have been as satisfactorily handled as it is possible to handle this commodity.

Loveland Factory

In the enlarged acreage to be put into beets this spring, there is sure to be some land infested with wild oats, and if this land has been fall plowed, the tendency is to work it up and plant the beets early in the spring, so as to get the beets up before the oats.

I have seen this plan tried a number of times, but the beets have always come out second best in the race, and usually have to be replanted.

The best plan is to work the surface early, get the oats sprouted and plow them under about the last week in April, prepare the seed bed and plant. Usually the ground is warm then and the beet seed will germinate quickly and the beets get big enough before the second crop of oats gets through, so they can be controlled with the cultivator.

Would suggest getting ready for beet help early and try and arrange to give the head of the family work on the farm until beet work starts.

* * *

Land that has to be plowed in the spring for beets should be turned over as early as possible and worked down so as to get the benefits of late frosts and snow, and be compacted and mellowed by seeding time. It is not as good as fall plowing, but much better than late spring plowing providing it is dry enough to plow early.

* * *

Now is a good time to see that the drill shoes are not worn out, and if they are dull it is economy to replace them with new.

* * *

When coarse manure is applied in either spring or fall, it should be disked in before being plowed under, as it will otherwise keep the land open, allowing the moisture to escape that would otherwise decompose the manure and carry its fertility to the roots of the growing crop.   —H. Scilley, Agricultural Superintendent.

Sterling Factory

An average of 20 inches of snow has fallen between December 18th, 1918, and February 20th, 1919. These conditions will give plenty of moisture until the spring rains start in.

* * *

Our fieldman at Sedgwick, (Mr. George Smith) reports tenants purchasing farms will lessen the tenants' percentage to about 70 per cent. This is the right step to secure better crops. About 20 per cent of the tenants will be required to move this spring, where formerly about 40 per cent were called on to do this.

* * *

Manure hauling is being done in most places in this district.

* * *

Mr. Johnnie Cook, president of the Great Western Commission Company, is putting out 100 acres of beets, which is his first venture in the beet business. Hurrah for Johnnie! It is now up to the fieldman to make Johnnie a successful beet grower.

* * *

Farmers will be notified when and where to call for their beet seed. Please be prompt on the right day, as some of our stations have no place to store any seed uncalled for, and the fieldman is required to be at other stations the next day. Seed will be shipped during the latter part of March.

* * *

If the farmer desires the Sugar Company to procure an Iron Age Sprayer for him, he should notify the Company at once in order to give them ample time to do this. The cost of this sprayer will be in the neighborhood of one hundred and eighty dollars, ($180.00) plus the freight charges. I cannot say just where you can obtain the Aspinwall sprayer this year. The Great Western Sugar Company has no sprayers on hand for sale.

* * *

In talking with farmers located in the Julesburg district, I find the general opinion to be that better beets are grown from May planting than by an earlier or later planting. This, perhaps may have been tried out in the earlier history of beets, in northeastern Colorado, where the land was not sufficiently rich. Consequently the beet was matured in early September, and was compelled to wait until digging time which would be on or about the first of October. This, I presume, was the reason for so much discouragement.

As to early planting. I remember in the fall of 1918, beets as a rule were not fully matured in northeastern Colorado. Now if you have a good, fertile piece of land, it will perhaps need more time to mature the beet, and as the prospects now look for the maximum acreage in the state, there is every indication that we will perhaps begin the harvest earlier. So prepare to begin with the others by having at least a small patch of early planted beets, especially if your ground is rich. Let us try early planting again, even though we may have to do this in a small way.

* * *

LOSSES ON BEET TOPS

Feeding beet tops in Sedgwick County in the past fall and early winter has suffered a set-back, due to improper methods, which it is to be hoped will benefit growers and feeders alike.

Regardless of the constant campaign for better feeding through the hauling in of beet tops, and the yarding of cattle, the general custom yet, is to over-run the fields for from forty to sixty days, and then throw the stock upon the market as top feeders. In 1917 these feeders brought fancy prices, and hence in 1918 the talk of yard feeding, by the Fieldman, was not productive of many results.

However, tops sold well, bringing from seven to ten dollars, the highest, by about twice, ever paid in this section. Hay was extremely scarce and no provision was made in most cases for storms which caught us. Cattle were thrown in about November 15th, feeders were purchased around ten cents, and a few cattle men bought tops for hold-over stock. The first month was ideal and apparently good gains were made. Then came the big storm, December 19th. Feeding hay was out of the question, and the feeders were rushed to market. From three to six days elapsed between the storm and the delivery of cars. At Sedgwick twenty-eight cars were marketed. Twenty-six of them reached Omaha weighing less than when purchased. Two cars were fed five tons of hay and showed a gain of 83 pounds per head.

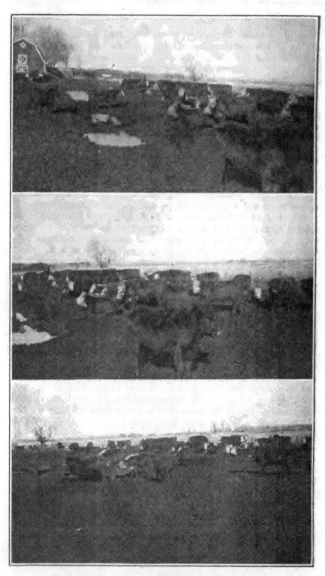

Cattle on Dry Pulp at Hershey, Lincoln County

Luckily the market was good and they sold for from $12.50 to $13.75, so there was small actual loss. After three weeks the remaining stock is cleaning up the fields, many of which were not more than 60% eaten up. The fields are suffering and the hold-over stock is going hungry or eating high priced hay with each succeeding storm.

Had even half of each field been stored in reserve, the gains could have been maintained and a handsome profit netted on the feeding.

Geo. A. Smith, Sedgwick District.

A firm seed bed is essential to a good stand of beets

Nebraska District

Pulp cattle are doing well and have made good gains. Hill & Company, of Scottsbluff, shipped six cars of cattle February 6th and received $16.75 and were on a low market. Going some!

* * *

It seems more attention is being given this winter to keeping the feed yards in shape and most feeders seem to be trying to keep the cattle on top of the ground. Why not note conditions of the yard now and plan some kind of an improvement to be worked out next summer, such as grading up the middle two or three feet, so cattle will have a dry place to rest on?

* * *

If you want to know anything about dried pulp, call on Mr. D. L. Parker of Scotsbluff, as he has made a study of it for years in California. Some cattle are on dried pulp and are not making good gains. Something wrong. Perhaps the ration is not balanced. Mr. Parker will be glad to help you work out a balanced ration. Cattle on dried pulp should make good gains. Many are, and others should do equally as well.

* * *

We are convinced that the North Platte Valley is the most beautiful place to live. California has nothing on us. No wonder that everyone who comes here decides to stay. Pay us a visit and be convinced.

* * *

We now have 40,711 acres signed. Everyone feels that the beet crop is the most profitable crop to raise; sure market, sure money, large profits, prosperity everywhere—that is the inevitable result of the beet industry.

We are living in a progressive age, so let everyone boost for good roads. They must and will come. Let us have a highway from O'Fallons, Nebraska to Lingle, Wyoming, for our share. Nothing does more to attract travelers than good roads. Good roads are also a great factor toward making a locality a pleasnt place in which to live.

Growers at O'Fallons are getting the "good roads bug," a mighty fine disease to have. O'Fallons growers never let anything go by that is good for the community and the way they go after things is not slow. If you want the details as to how they accomplish things, write Henry Fulk, Hershey, Nebraska.

A. H. Heldt, Agricultural Superintendent, Feb. 17, 1919

Lovell Factory

The Big Horn Basin has had a most splendid winter, and all things relating to winter work are in excellent condition.

* * *

The fattening of cattle, ewes and lambs has been very satisfactory, with many farmers contemplating getting their stock to the market before they had originally figured the same would be ready. There has been a small waste on all feeds, particularly the wet pulp.

* * *

Some fall work has been done by our farmers in this section, and it is a pleasure to note that some farmers already have their wheat in the ground. Others have worked down fall plowing and are cleaning up the farms preparatory to spring work.

* * *

A great deal of interest is taken in the handling of the grasshopper problem in all our districts, as considerable loss was had last year and the farmer realizes that concerted action alone will bring about the desired results.

* * *

Contracting is going along steadily, with many new growers who have taken the time to investigate the industry and the problems connected with the growing of the crop. The labor problem appears satisfactory, and we are glad to see that no anxiety should be felt on the part of our growers in this respect.

* * *

For the benefit of most sections of The Great Western's territory we wish to say that we still consider it necessary to irrigate before planting our beets, and that our spring work problems are somewhat increased in this respect, but we have the opportunity also of correcting and controlling our moisture conditions for the time of germinating. All that we wish to do is to remind our farmers to get after this crop early and have the main ditches and laterals in shape to deliver water for this land as early as possible so that the crop can go into the ground by March 1st.

Attention to the small details is necessary to the large tonnage of beets, or any other crop.

Missoula Factory

We have had a very fine winter for the stockmen here, but a poor prospect for late irrigation, as there is very little early snow in the mountains.

* * *

On account of a short crop of hay in the past season much of the stock in the district was sold to conserve hay. Little has been used up to this date and hay has dropped $8.00 to $10.00 per ton. Hay is now worth from $12.00 to $22.00 as against $20.00 to $30.00 per ton last fall, and it now looks as if a surplus in some districts will be carried over.

* * *

We are having some snow and zero weather at the present time, which ought to put the land in excellent condition for spring work. Winter wheat prospects looked very fine before this snow covering, and this will insure an excellent start this spring. Much spring wheat will be seeded over this entire district on account of the government guarantee.

* * *

Beet contracts are coming in here rather slowly, as farmers do not realize the value of a cultivated crop, not having had any in past years, and it takes a good deal of talk and persuasion to get a farmer to put part of his land in the best rotative crop grown, and one that we proved here last season would give better returns than any other in this district.

* * *

On contracts signed up to date we find only about 20% of the acreage fall plowed.

* * *

Fieldmen and self wish to tender our thanks and appreciation to Mr. Lippitt and The Great Western Sugar Company for the excellent business meeting and entertainment connected with it, and are busy at work with the idea that their best effort is none too good for the best company on earth.

—R. M. Barr, Agricultural Superintendent

Big Stocks of Wheat Found in Survey January 1st

Commercial stocks of wheat reported in a survey made by the Department of Agriculture for January 1, 1919, amounted to 191,775,-417 bushels. These holdings, reported by 9,303 firms—elevators, warehouses, grain mills, and wholesale dealers—were more than twice as large as the stocks held by the same firms a year earlier, the actual percentage being 215.8 per cent of the 1918 stocks. The figures refer to stocks actually reported, and do not represent the total commercial stocks of the country, nor do they include stocks on farms.

The commercial visibility supply figures, as published by the Chicago Board of Trade for the nearest date (Dec. 28, 1918) show 117,225,000 bushels of wheat, as against 18,963,000 bushels a year ago. Corresponding Bradstreet figures for January 4, 1919, show 129,627,000 bushels, against 26,476,000 bushels for 1918. As compared with the same date of last year, these figures, as well as those obtained by the more extensive survey, show a very great relative increase in commercial stocks of wheat on January 1, 1919.

Beets or Wheat

H. Mendelson

THE price of wheat and beets grown in 1919 is guaranteed by concerns amply able and willing to fulfill any financial obligations they undertake, namely Uncle Sam and the Sugar Company.

There are some differences, however.

All beets above 12% (and there is very little probability of producing beets with less sugar) will be paid for at the rate of $10.00 per ton. They will be received by the Company as soon as harvested. If beets are delivered before October first, the first payment will be by October 15th. With the big acreage in sight, the harvest will begin before October first unless the weather prevents. At any rate a large payment will be made by November 15th, and the rest will be paid for by December 15th.

The wheat price is set at $2.26 per bushel of number one hard at primary markets, making about $3.30 per 100 pounds at Colorado shipping points. Wheat of lower grade will be paid for accordingly.

A great deal of the spring wheat of the Defiance type, raised on our irrigated farms, will not be graded high.

A great deal of wheat raised last year was damaged by rains and graded far below the standard.

In other words, wheat will be paid for on a sliding scale with a guarantee only for the best grade. No minimum price is guaranteed.

When beets were paid for on a sliding scale, a minimum price was guaranteed.

Furthermore, the government so far has not guaranteed any definite time at which the crop is to be taken off the farmers' hands and paid for.

By November 29, 1918, out of the 917 million bushels of wheat harvested, about 329 million were still in the farmers' bins. It is well known that from time to time an embargo was levied on wheat shipments on account of congestion at the central receiving points.

This was partly due to the fact that ships were needed for troop and ammunition shipments. Also, the submarine retarded the trip across the ocean.

These causes of congestion of course will not exist during the coming year.

But the wheat crop of 1919 in all probability is going to be much larger than that of 1918.

The winter wheat harvested in 1917 was 17,832,000 acres.

The winter wheat harvested in 1918 was 24,406,000 acres.

The winter wheat planted for 1919 is 49,261,000 acres

So far, the average condition of the winter wheat all over the country is almost perfect, so that the acreage harvested will be also larger than that of 1918, and very much larger than the average harvested during the five years 1912-16 (about 34 million acres).

If the spring wheat acreage increases in the same proportion as the winter wheat acreage, there will be planted a total acreage of about 75,000,000 acres, as compared with an average of 53,000,000 harvested during 1912-16, and of 59,000,000 acres harvested in 1918.

This means, even at an average yield, a crop of unprecedented magnitude.

Meanwhile the European production will have increased and a large amount of Australian and some South American wheat will be available on the world's market, meaning in all probability that the world's market price of wheat, while our 1919 crop is being marketed, is going to be lower than the price guaranteed to the American farmer. This, of course, means that the government after buying the wheat from the farmer, will have to sell at a loss that part of the crop which cannot be marketed in the United States

The machinery by which this transaction is to be handled and the details of handling are not worked out at present.

It may be assumed then that there is no certainty about the date when the farmer's wheat crop will be taken and when it will be paid for.

Therefore, if you raise wheat you better provide some storage on your farm or at some elevator beforehand.

If you need the cash for your crop at a certain date, the beet crop will provide it.

The wheat crop may not.

Meteorological Report
FOR JANUARY, 1918 AND 1919

TEMPERATURES:	1919	1918
Mean Maximum	44.64°	35.29°
Mean Minimum	12.77°	3.35°
Monthly Mean	28.70°	19.35°
Departure from Normal	+3.10°	−6.09°
Maximum	56.00° on 23rd	71.00° on 3rd
Minimum	−20.00° on 1st	−24.00° on 10th
PRECIPITATION IN INCHES:		
To Date	0.00	0.34
For Month	0.00	0.34
Greatest in 24 hours	0.00	0.18 on 9th
Departure from normal	−0.36	−0.04
NUMBER OF DAYS:		
Clear	27	5
Partly Cloudy	3	17
Cloudy	1	9

STRAW IS VALUABLE

The other day we saw the remains of a straw stack that had been burned. The ashes were still smoking. We wondered if the man who had set fire to the straw really knew how much money he was losing by his act. No doubt, his one thought was to get the straw out of the way so that he might plow the field, and really he could have followed no better plan to accomplish this quickly and with little effort. The straw went up in smoke and was dissipated into the atmosphere, but with the smoke went valuable plant food and humus-forming material. As a consequence, the soil was just so much poorer than it was before and the coming generations have been robbed of just so much of their rightful heritage. We wonder if farmers think of this when they follow methods that tend to deplete the soil of its fertility. A straw spreader will help to put it back on the land where it can be used to help fertilize the soil.

From "Utah Farmer," Feb. 8, 1919.

Minnesota No. 13 Corn

A. C. Maxson

OULDER County can well be proud of its record in corn production, siloing, building and prizes won at the International Dry Land Farm Exhibit held at Denver in 1915, at the State Fair and the Stock Show.

The boys corn club of Boulder county has produced some very remarkable corn. At the Corn Show recently held at Boulder the quality of corn exhibited was much better than that exhibited at Longmont last winter. This shows that the farmers of Boulder county are learning what good corn is and learning to recognize the various varieties and to select typical ears of each.

In order that any locality may produce good corn the variety grown must be suited to the length of growing season prevailing in that locality. Much of Boulder county's success in corn growing is due to the work of County Agriculturist Simpson, who has selected Minnesota No. 13 Yellow Dent as the variety best adapted to the growing conditions of the county.

Boulder county has been placed on the map as a corn producing county through the efforts of the County Agriculturist, however, if it is to retain its reputation for producing good corn every grower in the county must pay strict attention to the selection of his seed corn.

Since Minnesota No. 13 is the corn most farmers are desirous of growing and since home grown seed is most suitable it is important that we know just what the characteristics of this corn are.

The writer just recently received a description of this corn from Prof. C. P. Bull of the Minnesota Agricultural College in which he describes Minnesota No. 13 corn as follows:

"The relative size of the ears of Minn. No. 13 corn varies somewhat with the latitude in which they are grown. In the central portion of Minnesota (this portion of Minnesota has a growing season about like that of Boulder Co., A. C. M.) the ears range from 8 to 8½ inches in length and from 6 to 6½ inches in circumference. The color chosen as standard for Minn. No. 13 is an old gold yellow, bright and of good lustre but not shiny. The space between the rows is a little over medium. The kernels in shape are the so-called wedge shape with 'square' shoulders at the cap and tip. They are about twice as long as broad and about twice as broad as thick. The depression conforms to the shape of the cap of the kernel and is only medium deep, what may be termed saucer shape. The cap of the kernels

Upper Left—Boys' Corn Club Exhibit   Main Exhibit, Boulder County   Upper Right—Main Exhibit Minn.
of Minn. No. 18   Corn Show   No. 18

should be very slightly wrinkled. The ears are cylindrical in form with medium sized butts, the shank scar being medium to small. The tip of the ear is generally not very· tapering and well filled in accordance with the season's conditions. The cob is relatively small and red in color. Typical ears have 16 rows."

There are many opinions regarding certain "show characters," and the value of a corn. This was evident in the corn show at Boulder. Roughness of kernel and small cob are considered desirable because they are many times associated with deep kernels which in turn are thought to be associated with high yields of corn.

This opinion influenced the judge who placed the prize on a very rough sample of Minn. No. 13 corn. In the accompanying cut

there are shown three types of roughness. "A" is too rough. This type is associated with late maturity and should be avoided in Boulder county. "C" is too smooth and is to be avoided because of low yielding qualities although this type is early maturing. "B" represents a type ear of Minn. No. 13 corn sent the writer by Prof. Bull of the Minnesota Agricultural college a short time ago. This is the type of kernel we should select in picking our seed corn.

Extreme depth of kernels and small cob do not necessarily mean large yields of shelled corn per acre. We often read that a certain corn shells 85 or 90% kernels. This means that 85 to 90% of the weight of the ear is kernels. When we compare this with a variety which shells but 70% kernels we are apt to be misled and consider the deep kerneled and small cobbed variety as best. This is not always true by any means. For example: We have a field of deep kerneled small cobbed dent corn and one of rainbow flint. The first we will say shells 85% and the latter 70%. The first has ears 9 inches long with 16 rows and the latter ears 15 inches long with 10 rows of kernels. Let us suppose that the kernels are exactly alike in size and shape in both varieties which means that there will be the same number of kernels per inch of row on the cob. Let us suppose again that there are 6 kernels per inch, then the dent will produce 864 kernels per ear and the rainbow flint 900 kernels per ear. Now if we have the same number of hills per acre in the case of both varieties and the same number of ears per hill, the rainbow dent will out yield the dent corn if the kernels each weigh the same as those of the dent corn or yield an equal amount with slightly smaller kernels.

This shows clearly that other factors, such as number of rows and length of cob are of greater importance than the per cent of corn shelled per ear. Again, depth of kernel as compared with cob diameter is misleading since the same depth of kernel on a 10 rowed ear looks much nicer from the show point of view than it does on a 22 rowed ear.

Thickness of kernel and width of kernels on the same sized cob are important. A cob 9 inches long and 3 inches in circumference bearing kernels with well filled shoulders will shell more weight of corn the thicker the kernels are. This is true because the space between kernels, however slight it may be, does not represent corn. If the kernels are double the thickness, the spaces will be reduced to one half in number and the weight of corn correspondingly increased.

This all shows the importance of selecting carefully the seed ears, not from the point of roughness, smoothness, size of cob, depth of kernels, length of ear or number of rows separately but with due regard to all these points.

Avoid too rough or too smooth kernels and too long an ear in Boulder county.

Let us pay more attention to the selection of our seed corn this year than ever before.

No amount of after cultivation can make up for poor cultivation

GENERAL OFFICE BUILDING

FACTORY
BRIGHTON-COLORADO

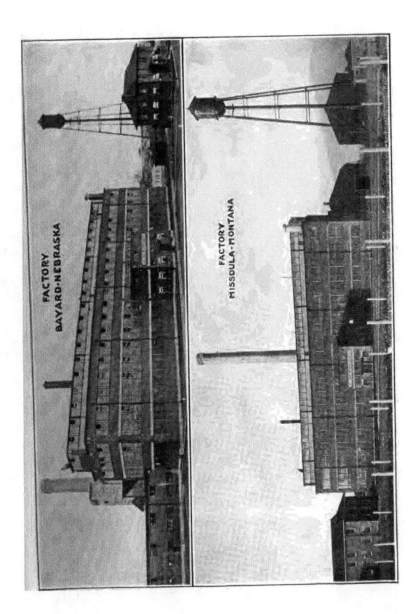

FACTORY
BAYARD·NEBRASKA

FACTORY
MISSOULA·MONTANA

FACTORY
SCOTTSBLUFF–NEBRASKA

FACTORY
LOVELAND - COLORADO

FACTORY
LONGMONT - COLORADO

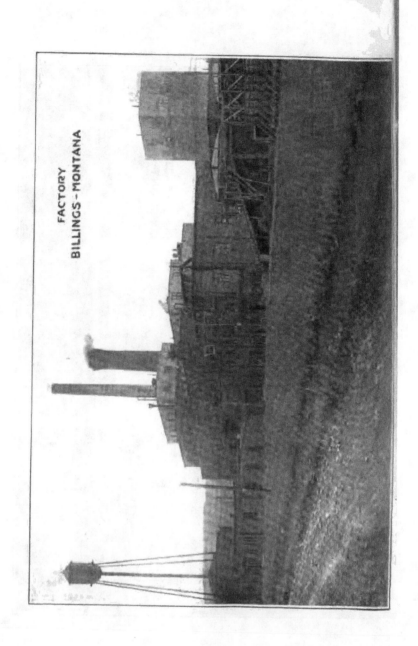

FACTORY
BILLINGS - MONTANA

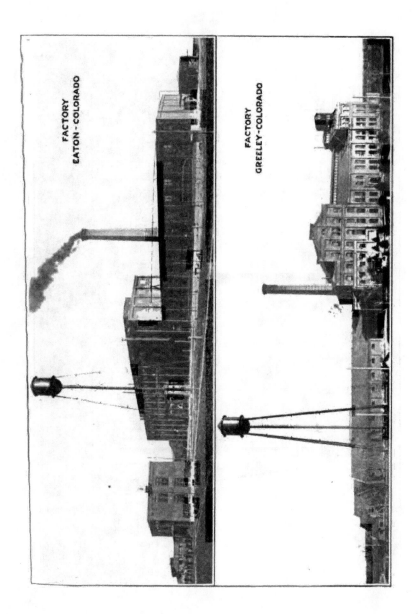

FACTORY
EATON - COLORADO

FACTORY
GREELEY-COLORADO

FACTORY
LOVELL—WYOMING

FACTORY
STERLING–COLORADO

FACTORY
BRUSH–COLORADO

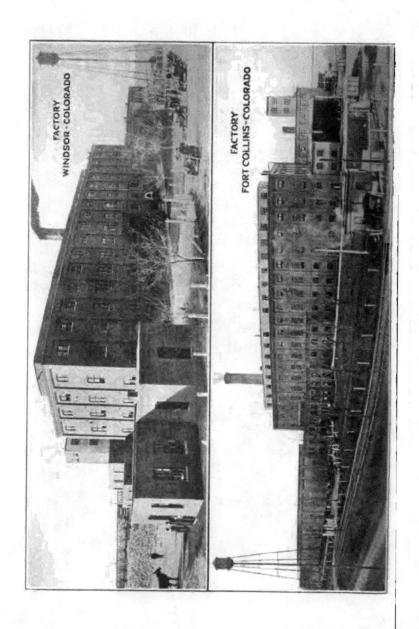

FACTORY
WINDSOR · COLORADO

FACTORY
FORT COLLINS · COLORADO

FACTORY
FORT MORGAN·COLORADO

FACTORY
GERING·NEBRASKA

This building houses: (1) Bag Factory; (2) Research Laboratory; (3) Pattern Shop; (4) Storage.

Address Given at Annual Meeting of
Boulder County Farm Bureau

C. L. Hover

I F you were a partner in a concern in which you had $5,000.00 or $10,000.00 or $25,000.00 invested, would you show enough interest in the annual meeting of the partners to attend? Certainly! Would you take an interest in the proceedings of the meeting? Undoubtedly! Would you be concerned about the development of the business? Assuredly! Now as a matter of fact, ladies and gentlemen, this is precisely the present situation. The concern is the Boulder County Farm and you and I and a lot of others are partners in the undertaking, and each of us has from $5,000.00 to $25,000.00 and perhaps $50,000.00 invested in it.

Today we are gathered here to discuss its development. This business was started back in the 50's and from that time to the present it has operated without interruption with periods of prosperity and times of adversity. In its 60 years of existence in has made marked progress. It is now a going concern with a bright future, and in all probability it will maintain a good record as long as civilization exists in Boulder county.

In skimming over the past history of this business in which we are all partners, we can distinguish four well defined periods, three of which it has passed through and the fourth it is now passing through.

The first period can appropriately be referred to as the "virgin soil period." This was when men first came into the county, broke up the sod and cropped with small regard for replenishing the fertility of the soil. At first they prospered, but the critical time came when the soil failed to respond as usual to the constant demands of the partners, and the business showed gross receipts of 35 bushels per acre falling to 15 bushels per acre. Observe that during this period we had no general manager but the business was being run by the partners acting independently.

In the fullness of time by accident someone introduced the alfalfa plant, and it proved so well adapted to soil, climate and altitude that it soon became a general crop. Fortunately, being a legume, its use restored the fertility of the soil and gross proceeds again rose to 35 bushels of wheat per acre. We can reasonably refer to the time of the introduction of alfalfa as the beginning of the second period of our business and term it the "alfalfa period."

This period experienced its critical time also for hay accumulated and was hard to sell. Soon the hay baler put in its appearance followed by the alfalfa mills, and the result was that our hay was shipped to Kansas and Nebraska and with it the fertility of our lands to build up the farms in those states at the expense of this business of ours. Let me pause here long enough to call your attention to the fact that the Boulder County Farm Bureau was organized to protect us from just such mistakes as this. We were at that time still without a general manager. Our gross proceeds again fell off, and another critical period was developed.

During this period, as a cultivated crop, potato growing came into practice for some of the partners had realized that our stiff clay soil craved and demanded a cultivated crop in the rotation of crops and would not respond well without it. Potato prices however, were always uncertain and after a few years the blight made its appearance. Fortunately at this crisis there came men who had faith in the sugar beet, and, just at the moment when the potato crop was failing us, the beet crop, requiring still more intensive cultivation was introduced. Soon after this our potato cellars caved in and we were on a beet basis. With the introduction of the beet our lands doubled in value, our towns and cities doubled their population, and this can justly be termed the "beet period" in the history of our business.

Again gross proceeds increased and the business thrived, but we were still without a general manager—the Farm Bureau not having been formed—and the business again drifted into financial difficulties. Of all crops the beet crop needed a fertilizer. The hay balers were still running at large. The alfalfa mills were still sending the fertility of Boulder county over into Kansas and Nebraska, and this bad practice was aided by the sugar factory when it undertook to feed the by-products of the factory at the factory instead of on the farms. The by-products were, in themselves, not a balanced ration and hay was needed so between the hay balers, the alfalfa mills and the sugar factory, our business, which was still without a manager, was being skinned. Our commercial associations saw the danger and gave the alarm. The sugar factory also soon saw the point and made arrangements to have the by-products fed on our farms. But the sight of a great business, in which hundreds of thousands, yes millions of dollars were invested and thousands of people employed, running without a head attracted the attention of the County Grange and the commercial associations and at this period steps were taken to organize the Farm Bureau.

The silo had already made its appearance, but with the employment of a general manager, the Farm Bureau had now been established and a county agent secured, silo building went ahead with leaps and bounds. The hay balers grew fewer. Alfalfa mills ran short on raw material. The by-products of the sugar factory found their way into our business. Cows, steers, sheep and hogs multiplied, and at last we entered into the fourth period in the history of our business, the period of the establishment of natural fertilizer factories on our farms which we can refer to more briefly as the "live-stock period."

It is this period through which this business of ours is now passing, and we may well congratulate ourselves upon its present strong position. It has established the finest forage crop known, namely, alfalfa. It has developed two cultivated crops, namely corn and the intensive crop known as the sugar beet crop. It has increased the gross proceeds from the king of grain crops, namely wheat. Through its general manager it is establishing the best known varieties of three of its main crops, namely Grimm alfalfa, Minnesota No. 13 corn and Marquis, Defiance and Turkey Red Wheat. Lastly it is establishing all over its broad acres natural fertilizer factories which with wise rotation of crops promises a wonderful development in the years ahead of us.

It may be that this will be the last distinctive period for many years, the "live-stock period." With a managing head, the Farm Bureau, well established, we have reduced to the minimum the possibility of another critical period creeping into our establishment unannounced.

If you will turn to the chart on the wall, you will notice that one of the projects that the management will undertake this year is that of Live-stock improvement. It has a direct bearing on the development of this period, the "live-stock period." The other projects, namely crop improvement, farm business and home demonstration, aim at improving various processes on the farm inasmuch as under them, problems are constantly arising and as constantly calling for rational solutions.

In conclusion let it be borne home that the duty of your general manager—the Farm Bureau—is to encourage in every sensible and effective way good farming and good farm management. And why? Good farming and good farm management always pay, no matter what the prices of farm products are; poor farming and poor farm management never pay no matter what the prices of farm products are. And for these reasons, good farming and good farm management are the underlying fundamental requirements in any successful agricultural development and therefore the fundamental requirements in the development of this business in which you and I and scores of others are engaged as partners.

Spacing Test

J. F. Jarrell

T HE danger of losing tonnage on a beet crop by spacing the beets too far apart is not understood by the average farmer. Most farmers plant their beets in rows twenty inches apart and would probably consider it too revolutionary to plant them in rows of twenty-two or twenty-four inches apart for fear of losing tonnage. This same farmer will often allow his beet labor to block his beets to suit itself and not offer a word of objection or even know in some cases that he is being imposed upon.

All of us have seen beet fields where the thin stand was the cause of low tonnage, but few of us have seen fields where too thick a stand has caused low tonnage and we are not very apt to see such a field for the hand labor usually attends to making the stand as thin as the farmer will permit because a thin stand is easier to harvest than a thick one. To offset this some farmers have offered their labor a bonus of a certain amount per ton for every ton above a stated yield. Thus: The farmer may agree to pay the usual contract price of $25.00 per acre for a fifteen ton crop and 50 cents per ton additional for every ton over fifteen tons per acre. Good labor on good land is always ready to enter into such an agreement and it will pay both the farmer and the labor. The objection to such a plan seems to be in the case of poor land and poor labor.

There being no effective means of handling this question, it settles down to a farmer's problem, and it is up to him to see that when his thinning is being done that the labor leaves as good a stand as is possible under conditions as they exist on his farm. Now we wonder what is a good stand? And how far apart should beets be spaced to secure the best results?

The following table will give an idea of spacing and the number of beets left per acre when beets are spaced to average as shown, considering the rows to be 20 inches apart.

Average Spacing	Sq. in. per Beet	Beets per Acre	Tons per A. if Beets Av. 1½ lbs. each
12 inches	240	26136	19.602
14 inches	280	22400	16.800
16 inches	320	19600	14.700
18 inches	360	17424	13.068
20 inches	400	15681	11.760
22 inches	440	14256	10.692
24 inches	480	13068	9.801

From a theoretical standpoint the above table seems conclusive in its showing of a decreased tonnage as the square inches per beet increases above 240, yet there are other ways to look at the thing which tend to modify this showing. The size of the beet of course will increase some as the spacing increases but experience tells us that 240 to 320 square inches per beet is the best under conditions as they exist in northern Colorado. Of course a farmer with exceptionally rich land will grow his beets closer than the farmer who has poor land. Then, too, a plentiful supply of irrigation water may develop to the maximum yield a thicker stand than a deficient supply of water would. The farmer has to consider many influencing factors which enter in, to cause the value of his crop to fluctuate.

The writer realizes as well as anyone the many difficulties under which a farmer has to labor to secure the best results from the hand work done on his beet crop. A careless cultivator man will sometimes cause considerable loss by cutting out the beets with the cultivator. We can always tell where he went to sleep by the open spaces affecting four rows alike here and there in the field. A careless irrigator often causes loss of stand by needless digging, dyking and flooding.

There is of course no positive rule to follow and if there was, we would not follow it. So it is up to the fellow who is most interested (the farmer) to see that his crop is properly handled.

Frequent visits to the field from the time the seed is planted until every beet is on the wagon to go to the dump will pay any man who has the inclination to secure all he can from his beet crop, and if any one time is more important than all other times it is when his beets are being spaced in the row.

FARM MANURE

Some farmers have learned the value of manure; the help it is to the soil, what it will do to increase the crops. Manure is the greatest available resource for maintaining soil fertility.

Recently, when discussing why a certain man had such a good yield of sugar beets, one man said: "Why that fellow hauls manure all winter—from any place he can get it." When this farmer was questioned as to whether it paid him to spend his time hauling manure, he was able to tell nearly just what increase each load of manure brought to him in the way of increased tonnage. What our soils need is humus. Soil experts tell us we have plenty of lime in our land, but what we need is plenty of good manure.

A great deal more manure could be 'made' for every farm, if a little study was made of how to do it. Take the question of liquid manure; if you do not have a concrete vat to hold it, use plenty of bedding such as straw and let this soak it up and it will answer the same purpose to a great extent. The important problem for every farm is fertility, and plenty of farm manure will be the greatest help.   From "Utah Farmer, Feb. 8, 1919.

Why Not 100 Per Cent on Smut Treatment

H. H. Simpson

I T seems strange that after so much has been written and said about treating seed to prevent smut, that some farmers will continue to sow untreated seed, even when they know a loss will follow. Last year there was more smutted grain in this section than for many years. Many farmers did not realize their grain was smutty until they threshed or took it to the mill. The writer inspected many fields during the harvest. season and very few were found to be free. The percentage of smutted heads ranged from nothing up to 15 per cent. If every farmer had treated his seed, Boulder County alone would have had ten to twenty thous-· and more bushels of wheat, to say nothing of the money which would have been saved for the farmer due to dockage.

Here is one illustration of how the loss totaled up. While watching a certain threshing outfit at work I noticed the presence of smut. The farmer showed me his elevator checks and told me he was being docked 30 cents per hundred for smut. Now basing his yield on, say, the average of 30 bushels per acre, this dockage alone meant a loss of $5.40 per acre. Estimating that he must have had at least 10 per cent smut heads in the field, there was an additional loss of three bushels per acre in yield, worth $5.72, meaning that the farmer suffered a total loss of $11.12 per acre on his wheat crop because he did not treat the seed. On a 20 acre field this would be over two hundred dollars, which would pay for a lot of treating, especially when we know the actual cost for treating is less than five cents per acre.

Of course we have farmers who make the statement that they treated their grain and noticed no difference; the heads smutted just the same and sometimes the untreated wheat did not smut. These arguments prove nothing. The effectiveness of the formaldehyde treatment has been proven, and where it fails to prevent smut or .at least cut it to a minimum, it is usually because some mistake was made. Where an untreated field shows no smut, it is simply a fortunate circumstance and no proof of the weakness of the treatment.

One mistake frequently made is to treat seed that contains smut balls. This is useless, for the solution does not penetrate the balls, which afterward break and the unharmed spores re-infect the seed. If smutted grain is to be treated, it should first be cleaned and the smut balls fanned out. Smut in winter wheat may sometimes show up due to the soil being badly infected while threshing smutted grain.

The treatment should not stop with wheat, but oats and barley should also be treated. In using the formaldehyde treatment, add to 40 gallons water one pint formaldehyde, and either the sprinkling or dipping method may be used. If you prefer to dip, put the grain in sacks, dip and leave in the solution ten minutes, then remove and continue to dip the other sacks until all have been treated. Now pile the sacks together and throw an old blanket or canvas over the pile and leave not over twelve hours. Remove the canvas and immediately spread the grain out on the canvas to dry, unless it is to put in the drill for planting. If you choose the sprinkle method, put the grain in a light wagon box, on a good floor, or on a canvas. Sprinkle the grain as it is being shoveled over. Cover, the same as the dip method and leave for twelve hours and then spread out to dry.

Why can't we make Boulder County 100 per cent, and everybody treat all their grain this spring?

A firm seed bed is essential to a good stand of beets

JOFFRE
Two-year-old Registered Percheron Stallion, owned by
Wm. Hanson, Longmont

165

As Early as You Can—

Make up your mind what you are going to do.

Do it.

Have harness, plows, levelers, drills, in perfect working order before the spring opens.

Get your horses fed up so they can stand the work.

Spread whatever manure you have.

Roll cloddy fields.

Double disk before plowing.

Harrow at least twice, what you plow each day.

Most farmers plan to raise a bigger beet acreage this year than they have had for a long time.

This means the harvest will start earlier than during the last two years.

Therefore—

You Better Plant as Early as is Advisable in Your Locality

Are Your Grain Troubles Answered?

1. Question

Is formaldehyde as good (effective) in the treatment of cereal smuts as copper sulfate (blue vitriol)?

Answer.

Formaldehyde is better than copper sulfate for treating grain. Formaldehyde costs 1-6 as much as copper sulfate and experiments have shown that the seed are less apt to be injured by formaldehyde.

2. Question.

What strength formaldehyde should be used?

Answer.

Forty per cent formaldehyde solution procured at any drug store at the rate of one pint to 40 gallons of water.

3. Question.

Is the method of sprinkling formaldehyde on the grain effective?

Answer.

If the grain has been thoroughly cleaned in a fan mill, the sprinkle method is all right. However, if your grain contains smut balls, which can be removed by a fan mill, it is a waste of time to use the sprinkle method. If you do not have a fan mill, use the open tank method and skim off the smut balls.

4. Question

I treated with formaldehyde, and had just as much smut as I had the year before, when I did not treat at all. How do you account for this?

Answer.

This is a question which is commonly asked. Perhaps the grain was not covered for the required time after treatment or perhaps it contained smut balls which were not removed before treating, or maybe the grain was not dipped for long enough time. (Write Extension Division of the College for information on "Treatment of Smuts.")

5. Question.

Will formaldehyde control all smuts?

Answer.

No. Corn smut, and loose smut of wheat and barley are not controlled by formaldehyde treatments. The reason why these smuts are not controlled by formaldehyde are: The corn smut organism lives over in the soil and the loose smut organisms of wheat and barley live inside the grain, so in either case, disinfection with formaldehyde is of no avail.

—James Geodkin, Colorado Agricultural College, Fort Collins Colo.

Be Prepared for The Web Worm

W. S. Henderson

Last year there was a great loss caused by web worms, undoubtedly the most severe in the history of the county. It is my opinion that the greatest precaution possible should be taken for the coming season. Although some think we will not be troubled with the web worms this wear, it is always best to be prepared and to use every method to prevent the spread of any pest.

We know the web worms are in the ground by the millions, but what percentage may hatch, none can tell; so the thing for the farmer to do is to fight them before they start to fight him.

The first brood is usually hatched in weeds along ditch banks, in the ends of fields and in weedy waste ground. The worms are only buried from three to four inches in the ground and the working of the fields will kill all the worms in the ground worked.

Get out early; burn the weeds; and disc the ditch banks, all weedy patches and waste ground. The worms breed principally in patches of Russian thistle and lambs quarter.

In this connection, the value of early discing in the spring before plowing may be emphasized. This should be done as soon as the frost is out of the ground. The discing and stirring up of the ground will cause the worms to be exposed to freezing and the hot sun. The disc itself destroys many worms. These same methods will apply to fighting the grasshopper pest.

Where the main ditch banks are disced it would be well to take a V-ditcher and clean the ditch banks at the same time. This will destroy any worms that are in the sides or bottom.

It will be easier and cheaper to fight web worms in the ground than in the beet fields. While this method of combating the web worms may not kill them all, it will, no doubt, get the majority.

—From "Morgan County Farm Bulletin," Feb. 15, 1919.

169

About Feeding Beet Top Silage

 OME of the daily papers give an account telling how horses are dying from eating beet top ensilage. We do not know any of the details and will not make any comments as to why the horses may have died.

There are a few suggestions, however, that are very important to those who may be feeding ensilage of any kind. If silage is allowed to remain exposed to the air for some time, it will cause a mould to grow and in this condition it should not be fed to any kind of livestock. In feeding corn or other ensilage from a silo, enough should be fed each day in order to keep the top fresh and free from any mould. In feeding beet top silage, uncover only a small part of the silo at a time, what you will feed up within a day or two. Never allow your stock to help themselves from the silo. Farmers who are successful feeding beet top ensilage feed about three pounds per day for lambs and for larger sheep about four pounds. This is always fed with some kind of roughage like alfalfa or other kind of hay. Cattle can be fed from 25 to 40 pounds depending upon the size and condition. For dairy cows the common practice is about 25 pounds. Silage of any kind should be used as only a part of the feed. Feed it to horses in very small quantities. If other feed is available it is better not to feed any kind of silage to them.

There are thousands of tons of beet tops grown every year in this state and the feeding of them should not be condemned because some animal may die from over eating of beet top silage or eating mouldy silage. Last year a number of livestock died from eating mouldy straw but this is no reason why good, clean straw should not be used as a feed for livestock.

If a horse was permitted to eat all the oats he could, chances are the animal would be sick and it might cause its death.

Beet top and corn silage should only be used as part of the ration for feeding livestock. Many of our successful feeders are using beet top silage this year with good success. Care, however, should be exercised in the amount fed and never feed spoiled or moulded silage.

Prof. John T. Caine III of the Department of Agriculture was asked about feeding beet top and corn silage and made the following statement:

"We do not recommend beet top silage for horses except in small quantities nor do we recommend it for small calves. Most of the trouble probably comes from mould, caused by the silage being exposed to the air too long. If they feed it fresh and feed it in small quantities, we think it can be done without any loss. We recommend it, however, as feed for dairy cows, cattle and sheep. The same is true in regard to corn silage and it should be fed in limited quantities to horses and should not be fed to calves until they are several months old.

170

The following letter is from Hyrum Timothy, agricultural superintendent at Greeley, Colorado, for the Great Western Sugar Company, when asked if beet tops were injurious to cattle:

Greeley, Colorado, Dec. 17, 1918.

Dear Sir,

I beg to acknowledge receipt of your letter of December 13, 1918, in which you ask in regard to siloed beet tops being injurious to sheep and cows. In answer will say that we have fed siloed beet tops for the last eight years and have not lost one single cow or sheep from the effects of feeding siloed beet tops. All our veterinaries are very frank in stating that siloed beet tops are good wholesome feed for sheep or cattle and is very good feed for fattening such stock, and recommend it to all farmers.

It might be true that some farmers have lost some cows or sheep while feeding siloed beet tops, but our veterinaries, after a careful examination, say that the deaths did not occur by reason of feeding siloed beet tops. Our farmers who have siloed beet tops say that a ton of properly siloed beet tops, properly fed, will put as much flesh or fat on a cow or sheep as a ton of well cured alfalfa hay, and those who have fed for years will say so.

I do not know from what standpoint a farmer can say that siloed beet tops are injurious to sheep and cows. It is true sometimes sheep and cattle die when fed on corn and the best of hay; but as a rule it isn't the hay or corn that kills the animals; it is some other cause.

We have lost a few horses in northern Colorado by pasturing beet tops, where the plow missed a good many beets; and especially in sandy soil the horse in eating the beets in the ground swallows so much sand that the sand clogs the stomach and kills the horse. We have taken as much as thirty pounds of clean sand out of the stomach of one horse, as the stomach will not digest the sand. There is another way you can kill your horses very readily; by turning work horses that have worked steadily and have been fed on other food into a beet field where the tops are frozen or take the horse up from frozen beet tops and start him off too fast or work him too hard. This will cause a fermentation which will soon turn to blood poisoning. This is what the veterinaries have found to be true. But, they do say if you are careful and do not feed too much frozen tops to horses that are unaccustomed to frozen tops, and do not run or overwork the horses that the frozen tops will not hurt them, with the exception of very rare cases.

The above is our experience in feeding siloed beet tops and also frozen tops. You could not make any of our farmers believe that siloed tops will kill or damage stock, and our veterinaries say it is impossible for siloed tops to be injurious to cattle or sheep, as they are a splendid ration for both sheep and cattle.

Yours·very truly,

H. Timothy,

—From "Utah Farmer."   Agricultural Superintendent.

Representative Live Stock Sales at Denver Yards

This table is based on representative sales as reported by the "Record Stockman." It is intended to show the trend of prices for the thirty day period preceding the day we go to press.

STEERS

	$14.00 to $14.45	$14.50 to $14.95	$15.00 to $15.45	$16.00 to $16.45
1000 to 1099 lbs.				
Jan. 26 to Feb. 10	64	....	203	80
Feb. 11 to Feb. 26	....	99	128	....
1100 to 1199 lbs.				
Jan. 26 to Feb. 10	....	16	1	42
Feb. 11 to Feb. 26	23	48	70	47

LAMBS, FREIGHT PAID TO RIVER

	$15.50 to $15.95	$16.00 to $16.45	$16.50 to $16.95	$17.00 to $17.45
70 lbs. and up				
Jan. 26 to Feb. 10	270	477	....	....
80 to 89 lbs.				
Feb. 11 to Feb. 26	533	....	19	5969
90 to 99 lbs.				
Feb. 11 to Feb. 26	....	434	459	500
100 lbs. and over				
Feb. 11 to Feb. 26	....	446	....	....

STEERS

In the first period of two weeks, thousand pound steers ranged from $14.00 to $16.50 with an averagee at about $15.00 to $15.50.

During the second period of two weeks, thousand pound steers ranged from $14.50 to $15.50.

Heavier stock went as high as $16.50, but few representative sales were recorded.

LAMBS, F. P. R.

In the first period of two weeks, stock weighing seventy pounds

172

and up ranged from $14.50 to $16.00, with $15.50 probably as an average.

In the second period of two weeks, eighty pound stock predominated at $17.00 to $17.50.

HOGS

Two carloads of hogs grown by the American Beet Sugar Company at Rocky Ford recently topped the market at $17.20, the highest price paid in many weeks. The hogs were pulp fed and averaged 243 pounds.

I Just Broke Even Before—Now I Make $5,000 a Year

Victor Garvin

WHEN I first came to Colorado and undertook to raise sugar beets, I at first followed much the same plan under which I had been brought up. I raised only beets on my 40-acre farm.

Previous to my entering the business of beet raising, I had lived in one of the dry farming sections of Eastern Oregon, where nothing but wheat was raised. Although slight diversification and the addition of more live stock was possible, yet we preferred to gamble with wheat and run the chance of getting one good crop every third year. We did not stabilize our type of farming by diversification and the addition of live stock, because then it would not be so easy to get away from the farm and spend our hard-earned money for a winter vacation.

At first, from my beet farming, I made but a bare living and just managed to break even. Although with my system of farming, I was able to get but an average of 8 tons of beets to the acre, yet around me I saw many instances of men who averaged as high as 25 tons, and had better improved places, better homes and other comforts. Time and again I asked myself why it was. There must be some basic principle involved which I had overlooked.

Realizing that if I could solve the mystery it would mean hundreds of dollars to me, I sat down one night, took a pencil and noted some of the many differences between my type of farming and that of the numerous successful beet raisers around me. I shall outline them, as well as the system of management which I adopted, in the hope that it may be of help to some other farmers.

In the first place I found that, whereas I raised but the one crop—beets—the successful farmers not only diversified, but rotated

173

their crops. Most of them started with alfalfa. Some of them planted it with a nurse crop of grain, others drilled it alone. But at least, they had the alfalfa, which added not only organic matter but likewise nitrogen to the soil. For two years or more they would harvest hay crops and then plow under the third cutting of alfalfa, letting the ground remain in that rough condition through the w.nter. Then as early as possible in the spring, they would again plow the land, this time 10 to 12 inches deep, after which it would be disced and harrowed until there was a good, deep seed bed for the beets. I had so far followed the practice of spring plowing.

After having grown beets on the alfalfa land for one or two, and at the most, three years, they would plant a grain crop. Sometimes it might be winter wheat, drilled into the ground after the beet harvest. Again it might be spring wheat or oats, drilled after having plowed the ground. And from careful observation and inquiry, I found that a grain crop following beets was less weedy and gave a yield of from 25 to 50% more than if it had not been preceded by beets. Then the rotation was started over again, beginning with alfalfa.

Again, I found that whereas my farming operations ceased with my beet harvest, most of the successful growers engaged at least to some extent in live-stock feeding during the fall and winter months. They had their beet tops, hay and grain. Usually the only crop sold, besides beets, was grain. Sometimes and especially before the present high prices of grain, even that was fed. In addition to these feeds they would buy at a low cost beet pulp and molasses from the sugar factory.

Some fed cattle, others sheep, but here is the plan I found most of them followed, more or less closely: As soon as the beets were harvested, they would buy feeders and turn them in on their alfalfa and grain stubble, later pasturing them on the beet tops. In this way it was possible to get cheap gains and have the manure spread over the land. Then late in the fall they would put their stuff in the feed lot on a ration of beet pulp and alfalfa hay with molasses.

This system of marketing the products of their farms gave them better returns than had they sold at once, paid them well for their time spent, and above all not only maintained but increased the fertility of their land. That was why they raised larger crops of beets than I; that was why they could afford better homes, better teams, and automobiles to ride in. I had been taking the lazy mans's way.

So I adopted the general practices given above, with some modifications. I will briefly outline these with my results:

I became interested in the matter of beet tops and found upon investigation that although when pastured they return a fair profit, it was possible by siloing them not only to save what would otherwise be tramped into the ground, but actually to increase their feeding

174

value and at the same time have a very succulent feed for use during the winter or spring months. The agriculturist for the Sugar Company told me that in various parts of the country they were siloed in trenches and in the ground. So after my beet harvest had begun I dug a trench in a well-drained portion of one of my fields and near to my newly built feeding pens. This trench was made four feet deep with runways at each end, and just wide enough to accomodate a manure spreader.

Then, immediately after the beets were topped, I drove through the field with the manure spreader, having taken off the beater and loaded on the tops and crowns. Then the spreader was driven through the trench and the tops run off into it. The wheels of the spreader as well as the horses' feet packed the tops into the trench so as to exclude most of the air. The tops and crowns were alternated with layers of oat straw, there being a 12-inch layer of tops, then a 6-inch layer of straw. As nearly as I could estimate, I put in about 4 pounds of salt to each ton of tops. When the silo was full in fact rounded up well above the ground, I covered it with about 6 inches of straw and then an 18-inch covering of earth. A trench was made around the silo to catch any run-off water.

That year I had in only 15 acres of beets. I weighed the tops from an average acre and found I got 5 tons; or 75 tons of tops from the entire field. I had added about 6 tons of straw. Thus, in my silo, I had 81 tons when it was filled. Of course, the straw served to take up some of the excess moisture from the tops, still a considerable amount drained off, for by actual weight I took out but a little over 60 tons of feed. Now, according to the best authorities on feeds and feeding, as well as to a comparative chemical analysis of it, this silage is equal in feeding value to good corn silage. That fall corn silage was contracted for at from $10 to $12 a ton. Thus if we say only $10, my silage when fed was worth $607.50.

Purchasing Steers to Get Value from Feed

I purchased 40 feeder steers, averaging 954 pounds, on the Denver market, ran them on my roughage and then put them on my feed lot, on a ration of beet pulp, hay, and molasses. Later I began feeding my silage. The total time fed was 150 days.

This year, after having followed it for four years, I have increased my yield of beets from 9 to 17½ tons per acre and my yield of alfalfa from 5 to nearly 6 tons. I have enlarged my feeding operations and, with good luck, expect to make nearly $5,000 clear this year. Had I not changed my type of farming and adopted a better system of management I would still be lucky to break even. The changes I made are not difficult to follow, and in many cases can be applied to other than beet farming, I am quite sure.

—From "System on the Farm," February, 1919.

Department of Agriculture Offices In Denver

The following is a list of offices of the Department of Agriculture maintained at Denver, Colorado:

Forestry Service—

Headquarters for district comprising all forests in Colorado, nearly all those in Wyoming, two forests in South Dakota, two forests in Minnesota, one forest in Michigan and one forest in Nebraska —a total area of national forests comprising almost 22,500,000 acres. Smith Riley is district forester in charge. Four hundred employes are under his supervision. Offices 463 Post Office building.

Animal Industry

Field office in charge of Dr. W. E. Howe, 444 Post Office build-'ng; maintained for the investigation of diseases of animals in district comprising Colorado, Wyoming and Nebraska. Animal inspection service at Denver stockyards conducted by Dr. C. F. Payne, as inspector in charge. Under the latter is a force of forty-eight inspectors. Office at 303 Stock Exchange building.

Bureau of Public Roads—

302 Tramway building. J. A. Whitaker in charge of federal highway aid service as to roads, and H. C. Diesen, irrigation engineer, in charge of general experimental work regarding irrigation. Denver office now carries forty men. The federal highway aid territory of this bureau covers Wyoming, Colorado, New Mexico and Arizona; also Forest Districts Nos. 2 and 3 comprising the Denver and Albuquerque branches. In addition to this the irrigation investigations of this bureau cover Kansas and Nebraska.

Food and Drug Inspection Laboratory

518 Opera House block. R. S. Hilner, chemist in charge. A force of three chemists and inspectors gives attention to the enforcement of the pure food and drugs act in Colorado, Wyoming, Utah, New Mexico and a portion of Montana. Has jurisdiction over foodstuffs and beverages entering into interstate commerce.

Bureau of Markets—

Third floor, Custom House. Stuart L. Sweet, field agent in marketing and in charge of Denver Market station. The city is one of the most important outposts of the Market Bureau at Washington, which now maintains 2,400 employes in all its branches. Gathers and disseminates information concerning preservation and marketing of agricultural products for benefit of growers and dealers. During the heavy shipping seasons the office publishes daily bulletins on market prices, movements and conditions which are distributed free to farmers and dealers. Besides the bulletin and information work, six projects are carried on in Denver, which are:

Grain Standardization.—504 Cooper building, E. A. Hill, supervisor. Supervision of grain shipments and establishment of federal standards in District No. 31, comprising Colorado, most of Wyoming, eleven western counties in Kansas and portions of Arizona, Montana, Nebraska, New Mexico, South Dakota and Texas.

Seed reporting.—512 Cooper building, J. W. Dykes in charge. Investigates and reports upon seed prices, conditions and supplies.

City Marketing and Distribution.—Third Floor, Custom House, R. E. Brabil in charge. Makes daily investigations of farmers' market in Denver, issues bulletins to growers and publishes newspaper reports on garden truck.

Inspection of Fruits and Perishables.—Third floor, Custom House, J. N. Mosher in charge. At the request of growers and dealers, makes inspection of carload shipments of fruit and perishables on the tracks in Denver raidroad yards.

Transportation.—Third floor, Custom House, C. A. Heiber in charge. Makes a special study of car supply and movements for shipment of agricultural products in Colorado and surrounding states.

Live Stock Reporting Service.—214 Stock Yards building, R. E. Reynolds in charge. Issues daily reports on supply, demand and movement of live stock in the Denver yards and upon feeding and finishing of stock in Colorado.

Weather Bureau—

450 Post Office building. F. H. Brandenburg, forecaster; Frederick Brist, assistant forecaster. The Denver district office which maintains nine employes and has generl charge of 160 co-operative Colorado stations, and fifty river stations, covers in its jurisdiction Colorado, Utah, Arizona and New Mexico; also various important rivers which it is necessary to watch in order to send flood warnings. These include the Colorado river from its source to the Gulf of California, the Rio Grande from its source to El Paso, Texas, the Arkansas river to the Kansas line, the Canadian river in New Mexico to the Texas Panhandle, and the Pecos river in northeastern New Mexico into southwestern Texas. The office issues cold wave warnings to fruit growers and shippers and notifies farmers as to good alfalfa and potato harvest weather. The bureau also issues live stock, frost and fire weather warnings, maintains a printing department and publishes a monthly weather bulletin for the entire state of Colorado.

Biological Survey—

Second floor, Custom House building. L. B. Crawford, director in charge. Work consists of exterminating predatory animals and conserving of livestock in Colorado.

Bureau of Crop Estimates—

Second floor, Custom House building. W. W. Putnam, field agent in charge of Colorado district. Inspects growing crops in this region, furnishing figures upon which are issued official estimates from Washington.

Manure Wastage Largely Due to Lack of Handling Facilities

 year or more ago, just when every detail of both national and private life was being feverishly probed and investigated to check unnecessary waste and increase efficiency, so that we might to a greater degree help with the financing of the war and the feeding of our armies and our allies, and when soaring fertilizer prices seemed a barrier to the increased food production demanded of the farmers, Assistant Secretary of Agriculture Carl Vrooman made the astounding statement that the leaching of manure on American farms in a single year represents a loss of approximately a billion dollars! Although we have since become accustomed to speaking glibly of billions, particularly when discussing Liberty Loans, war losses and indemnities, yet the figure is far beyond our comprehension. In fact, much of the force of the statement is lost by its vastness.

We don't know just how Mr. Vrooman arrived at the total of his estimate. It is hardly probable that he had all of the farm manure piles counted and the time that each has been leaching computed. However, if you think his estimate is an excessive reflection on the business judgment of American farmers, just make a little survey and computation of your own. There are many standard works on the amount of stable manure produced by farm animals and on the loss in value manure suffers when exposed to the elements, which books will aid you in making your estimate.

County Agent G. N. Worden of Hancock County, Maine, recently made a careful survey of his county to determine the waste of fertilizer elements due to improper handling of farm manure, and his figures corroborate to a striking degree the estimate of Mr. Vrooman, assuming that Hancock County, Maine, may be taken as a fair average of the 2,936 agricultural counties of the nation. His estimate has the added advantage of being of comprehendible size and is expressed in terms which we can all grasp.

After the completion of his survey, Agent Worden addressed an earnest appeal to the farmers of Hancock county, which will no doubt be of interest to farmers of many other counties, and from which the following excerpt is taken:

"Successful crop production and good soil management go hand in hand. One thing is certain, and that is that unless a soil carries the food or fertilizing elements needed by crops, harvest time will never come into its own.

The principal elements of plant food required for plant growth are nitrogen, phosphoric acid and potash. We buy commercial fertilizers because they carry these elements, or are supposed to. This year we are paying approximately 40 cents, 8 cents and 30 cents for these elements respectively, and the same identical elements are found in the manure produced by all farm animals. This manure carries about 80 per cent of the fertilizing value of the food consumed, and when it is properly cared for, very little of its fertilizing value is lost.

"Under ordinary methods of care, however, here is what happens: In two months' time, manure thrown behind the barn where liquid portions leach away and the remainder is left to the mercy of the elements loses one-half of its value. In six months 60 per cent of its value is lost, and the most valuable and available elements go with the liquid portion, for this contains one-half the value of the nitrogen and two-thirds that of the potash.

"We find that the stock of the county produces manure valued at $796,427.82 per year, giving its plant food a value equivalent to what we pay for the same elements in commercial goods, and there is no other way to compute its value. Reckoning its value according to present methods of care, however, $437,257.77 of this possible value is unquestionably lost. With due respect to the few farmers who are actually preventing unnecessary waste by use of manure pits, sheds, etc., it would surely be fair to estimate an actual loss to the county of $400,000.00 per year, or nearly half a million dollars.

"Now, just what does this loss mean? It means the loss of money enough to buy 476 tons, or approximately sixteen 30-ton carloads of the highest grade commercial fertilizer. It would buy 15,385 tons, or approximately 513 30-ton carloads of acid phosphate. It would land at our county seat 63,191 tons, or approximately 2,106 30-ton carloads of limestone. It would pay the entire running expenses of Hancock county, as per budget of our county commissioners, for 1919 more than ten times over.

"If this isn't one of the most vital problems the county has to face, what is? Can the farmers afford to let this great loss continue, when it can be stopped, and have a right to expect profits from their business? Is the existence of a farm bureau justified that does not include in its program of work a campaign to reduce such a loss to the minimum? Most assuredly not!"

It is quite apparent that fertilizer elements of truly incomprehensible value are going to waste, and that the loss is quite general throughout most sections of the country. However, it is one thing to expatiate upon the marvelous transformation that could be worked by stopping this waste, and quite another matter to handle the millions of tons of manure which must be moved if this wonderful transformation is to be wrought.

Manure is the great by-product of the farm, but here in America, more especially during the last few years, farmers have been so dis-

179

tressingly busy with problems of the main products of the farm that many other things just had to be neglected. Furthermore, after a man who is unaccustomed to the work, pitches manure for a day or so, he ceases to wonder why it is that the conservation of manure on the farm is apparently so persistently neglected, while other work receives a fair degree of attention. Pitching manure from a soggy and odoriferous pile into a wagon and then scattering if from the wagon over a field is about the most disagreeable as well as one of the most laborious tasks of farming. As a means of disgusting boys with farm life it beats early-morning hand-milking two to one.

However, in the conservation of manure as in scores of other farming tasks, inventive genius has come to the relief of the farmer, and as a result we have the manure spreader—developed to a degree beyond all peradventure of an experiment, and demonstrated in thousands of cases to be an investment of indisputable economic worth.

The beds of most manure spreaders are considerably lower than the average farm-wagon bed, making loading decidedly easier than when a wagon is used. The unloading can be accomplished in one-third the time and at a saving of labor which cannot be computed. It

> No amount of after cultivation can make up for poor cultivation

is estimated that it will take 11 days to handle 100 loads of farm manure with a wagon while the same work can be accomplished in six days with a spreader.

However, the saving of man strength, good humor and time are not the only advantages to be gained in using spreader in the conservation of this great by-product of our farms. The saving of manure which can be effected with the machines is of almost equal worth. When the wagon is used it is hardly possible to do a fair job of spreading without using from eight to ten tons of manure per acre, and the more of liquid excreta there is in the mass the harder and more disagreeable the task. With a spreader an acre can be nicely covered with four or five tons, and recent experiments tend to prove that the latter amount gives most satisfactory results. Furthermore, Director C. E. Thorne of the Ohio Experimental Station, one of the greatest authorities in the country on fertilizer values, says that eight loads put on with a spreader are equal to twelve spread with a fork. This is due to the pulverization and more even application. Farm manure at the present time is worth from $5.00 to $6.00 per ton. Thus the saving in manure alone more than justifies the purchase of a spreader, to say nothing of the major considerations—saving time and labor and the stopping of loss through leaching of unprotected manure piles in barn lots. —American Farming

PERIODICAL

Through the Leaves

APRIL, 1919

THE
GREAT WESTERN
SUGAR COMPANY

THROUGH
THE
LEAVES

APRIL, 1919

Published Monthly By

THE GREAT WESTERN SUGAR CO.

DAILY CALL PRESS
LONGMONT, COLORADO
1919

TABLE OF CONTENTS

Snow on the watershed of the Big Thompson, March 1, 1919. Two feet of snow has fallen since this picture was taken.

Editors' Page

ITH the largest acreage of beets on record, some growers are wondering now about the delivery of their crop next fall. We want to assure you that we are putting all of our dumps in the best possible condition and we will make every possible effort at harvest time to help you get your crop delivered without delay. It is probable the harvest will have to start a little earlier than for some years, so the early planted beets are going to be a great help in the fall, as they will be the ones that have had a full length growing period in which to make tonnage and sugar. .

Now is the time when attention to the small details counts most. See that your seed bed is firm before planting. Many poor stands result from the seed being planted in loose ground.

WATCH YOUR DRILL. See that it is feeding continuously from all four openings, and examine the depth carefully. The small germs cannot force the tender shoot through a great depth of soil. One inch to 1½ inches should be the maximum.

A WORD REGARDING REPLANTING:

The grower often feels blue about a stand of beets and sometimes disks them up without consulting his fieldman. Uusually a 50 per cent stand of early beets is better than a 90 per cent stand of replanted beets. Do not destroy your first crop without calling your fieldman. Our rule is not to supply seed for replanting without the fieldman has approved.

LAND LEVELER   LONGMONT DIST.
Plan   Scale ½ inch = 1 foot

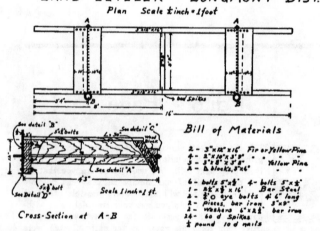

Bill of Materials

```
2 - 3"x12"x16'  Fir or Yellow Pine
4 - 2"x10"x3'9"     "    "    "
3 - 3"x8"x3'8"          Yellow Pine
2 - △ blocks, 3"x6"     "    "    "
6 - bolts 5"x½"   4- bolts 5"x½"
1 - 2½"x½"x16'   Bar Steel
2 - ½"⌀ eye bolts 4'6" long
2 - pieces, bar iron 5"x9"
2 - washers 6"x2½"  bar iron
24- 60 d Spikes
½ pound  10 d nails
```

Cross-Section at A-B

DETAILS of IRON WORK
Scale: 3 inches = 1 foot

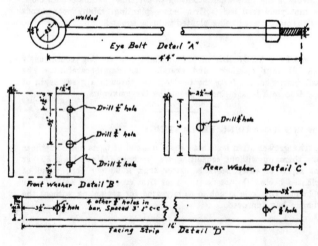

Eye Bolt   Detail "A"

Front Washer   Detail "B"

Rear Washer, Detail "C"

Facing Strip   Detail "D"

News from Factory and Field

Longmont Factory

LABOR FOR BEETS

From present indications there is going to be plenty of labor for all beets. The supply of Russian help is limited, but there are plenty of Mexicans.

* * *

In this connection we are finding that this year, more than ever before, the farmer with a good house for his beet help is getting the better quality. Just one instance occurred recently. Three different families (all good workers), who had worked at different times for a beet grower, each refused to consider this contract this year because of poor housing conditions. They all three signed other contracts. This matter of living quarters is uppermost in the minds of most of these laborers, and must be given consideration if you expect the better class of workers.

* * *

Twenty-five thousand acres of beets for Longmont is an increase of 100 per cent over a year ago. We hope this means 100 per cent more beets and not just paper acreage. We believe with the start so many are getting the results will be fine. Already a very large acreage has been planted and the seed-beds have been very well prepared. It is a pleasure to see the effort that is being made to get the beets planted early.

* * *

Some apprehension has been expressed regarding snow supply. Forest Supervisor H. N. Wheeler and his rangers report less snow now on the ground than normal; but he says, further, "Since the ground was so well saturated early in the season our water outlook should not be so discouraging as the actual depth of snow would indicate."

* * *

EMPTY SEED BAGS

Owing to the scarcity of bags for beet seed, we are desirous of having all bags returned when empty. In order to make it worth the grower's while to see that this is done, we are charging 50 cents each for the bags, in addition to the 10 cents per pound that is charged when they are weighed out with the seed. Credit will be given at the same rate for all bags returned in good usable condition. The only exception to this is Cotton Seed Meal Bags, for which the arbitrary 50-cents charge is not made, nor will this credit be given if returned.

Fort Collins Factory

EARLY ESSENTIALS FOR BEST TONNAGE

About a year ago we sent from this office a letter to the growers of beets calling to their attention the request from the Government for the greatest possible production to meet the needs of the world for food, and stated therein that, by careful and painstaking effort in regard to the work attached to beet growing early in the season, it was easily possible to increase the tonnage at least two tons per acre, which was equivalent, on our contracted acreage of last year, to setting aside from 1,500 to 2,000 acres for other crops. That is just what happened. Our average tonnage last year was two tons greater than the previous year and nearly that much larger than the average production.

We stated in that article that no other factor is so prolific in causing poor stands and in reducing tonnage as the crust that usually forms before the beets appear. We think that the results of last year demonstrate this fact to a greater degree than anything that has heretofore come to our notice. The weather conditions were such, and the moisture so well distributed, that practically no crusted conditions appeared. And to my mind the start that the young beets got by not being obliged to contend with a severe crust for existence is the factor that was responsible to a very large degree for this increased tonnage in the Fort Collins beet section, as well as in some surrounding territory. Of course, it cannot be said it was entirely due to this, for the interests are of too great magnitude to be determined by any single factor, but it is plain that this was the only condition that stood out strikingly that did not prevail in other seasons, when the result was not nearly so good.

We have frequently written and stated to the growers, and I hereby wish to emphasize it again, that the formation of a perfect seed beet and proper attention to the crust are the two things most essential to securing the very best results. To successfully meet this condition of a crusted soil, two things are primarily essential— first, a well prepared, firm seed bed, so that the drill in planting will make but a slight depression; second, a light harrow (especially constructed) at hand to combat the crust at the proper time. Without a properly prepared seed bed it is quite difficult to give the little plants relief with any implement whatsoever.

With a compact bed beneath by which the young rootlets can come in easy contact with the soil, and with the surface not crusted, allowing the leaves of the plants to easily and quickly get the sun and air, the crop is started under most favorable conditions. The weather conditions of the past year gave us this without effort on

188

our part. Breaking the crust with a light harrow approaches this natural condition better than any other way known. This means of relieving the crust is a most efficient cultivation. It relieves the danger of the small beets being destroyed by winds and frosts, and it puts the soil immediately in the row in fine condition for blocking and thinning. It further aids the bacterial action of the soil much to the benefit of the small plants. To some this may seem unimportant, but it is a very important consideration. In order to satisfy yourself in regard to this, break the crust on a portion of land as I have suggested, leaving some near by in the crusted condition, and notice the thrifty green color of the beets in the former as compared with the yellow appearance of those in the latter. Another advantage in the use of this implement is the rapidity with which the work may be done, which is often quite essential.

The crusted soil is usually attended by a more or less wet condition of that portion immediately underneath the crust, for which reason we should avoid the use of anything that has a tendency to weight or pack the land. I have often seen the soil harder where a roller had pressed the wet soil than it would have been had nothing been done. Nor is this wet land in fit condition to be even stirred with an ordinary cultivator. Furthermore, with the ordinary cultivator the work cannot be done rapidly enough to meet requirements, as frequently there is but a day or two intervening between the time when the work must be done to avoid injury and when the soil will permit getting upon it. With a harrow a large grower can get over his fields before the crust becomes too severe. All other implements are too slow.

The young seedlings should not be delayed in getting their leaves to the sun and air as quickly as possible, if the plant is to do its very best thereafter. If the young seedling plant is allowed to remain just under the crust for from one to many days, it becomes curled and is thus weakened, which has much the same effect upon the vitality of the plant and in the consequent yield of the crop as deep planting of the seed. There are but few instances where if a careful watch of the field is kept that the field may not be worked with this implement before any plants appear on the surface. There need be no fear of working the land until it is thoroughly mulched, as long as one can lift the crust without disturbing the seedlings. In fact, do not delay the work should a few plants appear, as you will destroy but few of them and relieve many.

Every beet grower should keep an especially careful watch of his beet field from the time the seed is planted until the plants attain four leaves, for this period is the most critical one with the crop.

It is not possible in an article of this kind to give instructions as to how each and every condition may be met in the best way, we can consider only those conditions that most generally prevail.

189

There are many instances that require modified methods, and when such is the case the fieldman should at once be consulted, for he is meeting these problems every day from many different angles, while the farmer sees them from but one.

The harrow for this purpose should be light and constructed so that the teeth may be tilted. The teeth so placed in the frame that every inch of top soil is broken. For small fields this harrow can be made very cheaply by taking three 2 x 4's twelve feet long and making a frame of them by placing them one foot apart. Into this frame may be driven 80 penny nails at an angle of 45 degrees, spaced six inches apart on the frame in such a way that the teeth cut the soil two inches apart. A more serviceable one consisting of two sections, each section having four hardwood bars 2 x 2½ inches, five feet long, placing ⅜ harrow teeth in them at intervals of seven inches, and alternating with each other so as to properly cut the soil, may be made by a blacksmith. These bars should be placed about nine inches apart and be held together with the same irons as an iron harrow, so that the harrow may be tilted at any angle according to the severity of the crust. This implement can be used for any purpose for which a harrow is needed and is often serviceable in preparing a good seed bed for crops of any kind.

I wish to especially urge all beet growers to provide themselves with such an implement, for taking into consideration the rapidity with which the work may be done, the assurance that no unnecessary packing will be given the soil, that the crust is lifted from the little plants rather than pressed upon them, that all the soil is cultivated in a most efficient manner, and that by the very nature of the work warmth and air are admitted to the soil rather than retarded, this implement is especially efficient and desirable for each beet grower. Samples of these harrows may be seen at the seed warehouse at the factory.

—H. H. Griffin.

At the recent dairy products show at Boise, Idaho, the Farmers' Brand of condensed milk put up at Loveland in this state was awarded the prize in competition with the Borden Condensery Company, the Carnation Milk Company, Utah Condensery Company, Libby, McNeil & Libby, the Colorado Condensery Company, the Maricopa Creamery Company of Utah and others. The Larimer County Condensery has grown into a most important industry, largely on account of the war, which proved a great help to all the condenseries in the country.

—From "Field and Farm," March 15, 1919.

Billings Factory

BILLINGS CLIMATALOGICAL DATA
January, 1919

		Date		
Maximum	67	23	Mean Maximum	48.5
Minimum	—5	2	Mean Minimum	20,9
Precipitation	.04	....		
Clear Days	16	...		
Part Cloudy	14	....		
Cloudy	1	....		

Frost Free Period Season 1918 in Billings Factory Territory

	Last frost in Spring	First frost in Fall		Frost free period
Billings	May 22	October	26	157 days
Huntley	May 22	October	7	138 days
Bridger	May 20	September	4	107 days
Miles City	May 31	October	29	151 days
Sidney	May 25	September	9	107 days

—C. S. MILHISER, Agricultural Superintendent.

Meteorological Report
FOR FEBRUARY, 1918 AND 1919

TEMPERATURES:	1919	1918
Mean Maximum	42.68°	50.50°
Mean Minimum	15.79°	16,37°
Monthly Mean	29.23°	33.44°
Departure from Normal	+1.23°	+5.44°
Maximum	60.00° on 17th	72.00° on 22nd
Minimum	2.00° on 24th	—6.00° on 27th
PRECIPITATION IN INCHES:		
To Date	0.53	1.55
For Month	0.53	1.21
Greatest in 24 hours	0.30 on 23rd	0.52 on 27th
Departure from normal	+0.36	+ 0.53
NUMBER OF DAYS:		
Clear	15	12
Partly Cloudy	6	14
Cloudy	7	2

A light harrow used extensively in the North Platte Valley for breaking crust and keeping the soil from blowing

The Farming Business and The War

H. Mendelson

The following figures are taken from the "Monthly Crop Reporter," published by the Department of Agriculture in Washington:

> If we call the average for the 5 years, 1909-1913 100
> the average values of plow lands in 1918 were 167
> the average farm wages in 1918 were 172
> the average crop prices in 1918 were 206
> the average live stock prices in 1918 were 211
> the prices paid for articles farmers buy in 1918 were 202
> the average yield per acre of all crops in 1918 was 100

This means that crop and livestock prices increased more than land prices, wages, and even prices paid for articles farmers buy, and that therefore farmers made more money than ever before. This is of course nothing new to our farmers. However, it means this—that our farmers are well prepared to weather a storm, if it should come. It also means that better farming will result. Good farming in many cases means sacrificing profits of today for the sake of returns in the future. There must be a reasonable financial reserve either in form of cash or credit before this can be done. It is just as true that good farming leads to profits as that good profits lead to good farming.

What the future holds nobody knows definitely. There is very little reason to believe that there will be a violent slump in grain or livestock prices in the near future if international commerce is restored.

Disinfect Seed Drill Before Using

The grain drill should be thoroughly disinfected before it is used for seeding purposes, according to James Godkin, of the Colorado Agricultural College, who says: "What is the use of treating grain for smut and then running it through a drill which may have millions of smut spores lining the box and other inside surfaces?

"The drill should first be cleaned out with a small broom or brush. It is then best to plug up the drill tubes, sprinkle the inside thoroughly with formaldehyde solution, close up the box and allow it to stand so that the fumes may come in contact with spores that may be adhering to the inside of the drill. Every precaution should be taken that the drill is entirely free of all the smut spores."

193

Ready for any emergency that may arise in the North Platte Valley.

Average Prices Actually Received by Producers in the United States

	Nov. 1 1918	Mar. 1 1919	
Wheat	206.0	208.0	cents per bushel
Corn	140.3	137.2	cents per bushel
Oats	68.2	62.6	cents per bushel
Barley	94.9	85.4	cents per bushel
Rye	152.6	132.2	cents per bushel
Potatoes	127.2	109.4	cents per bushel
Cotton	29.3	24.0	cents per pound
Butter	49.7	43.8	cents per pound
Hay	19.27	19.82	dollars per ton

The prices of all products declined, except those of wheat and hay. These figures represent averages for the whole United States. The price movement of some articles is not uniform all over the country. For instance, choice baled alfalfa was bought f. o. b. Denver, November 1st, for $26.00 per ton and March 1st for $21.00 per ton. No. 1 Timothy sold at both dates for $24.00 per ton.

No. 2 Hard Spring Marquis wheat also sold at the same price at both dates.

Choice baled alfalfa sold f. o. b. cars Kansas City, November 1st, for $31.50 to $32.00, March 1st for $32.00 to $32.50.

The price at Denver dropped $5.00 per ton, in Kansas City it remained about the same.

However, the average prices for the United States give a good picture of the general trend.

Since March 1st corn and oat prices have shown an increasing tendency which is likely to continue up to May or June.

195

Home of C. D. Turner, Showing Feeding Pens and Lambing Sheds at Rangeside Ranch in Morgan County, Colo. C. D. Turner is a beet grower, farmer, sheep feeder and breeder of pure-bred, registered Hampshire sheep. Fifteen head of pure-bred, registered Hampshire ewes are the foundation of stock.

Planting, When and How Deep

A. C. Maxson

Time to Plant

Few farming operations in connection with the growing of crops can be regulated by the calendar. The planting of beet seed is no exception. As is the case with the majority of the operations on the farm, there is a "just right time" to plant sugar beets in order to secure the best results in tons per acre. There is also a "too early" and "too late" date for best results.

The results in tons per acre are influenced by many factors, some within the control of the grower and others over which he has no influence. It is obvious therefore that if by manipulating the controlable factors a grower can, on the average, produce better yields a knowledge of the fact is desirable. One of the controlable factors which influences the yield is the time of planting. The proper time to plant varies with the climatic conditions and the locality. Favorable or unfavorable conditions of climate are indicated by certain readable out-of-door signs.

For instance, the presence of ice on our reservoirs indicates that atmospheric temperatures have not yet reached a temperature sufficiently high to heat the soil to a point where germination and growth will take place properly. Likewise, before the soil will be fit for the planting of beet seed all frost must have left the ground on the north sides of outbuildings and in other shaded places.

Since the ice on the reservoirs and the frost in shaded soil depends upon the severity of the winter, these may not be as safe indicators as the growth of alfalfa and the germination of weed seeds.

Beet seed will not germinate when the ground is too cool to germinate weed seeds or start a good growth of alfalfa, nor will beets make a healthy growth when weeds and alfalfa are retarded by cool weather. Therefore, regardless of the calendar beet seed should not be planted before wild plants begin to show signs of thrifty growth.

Over a period of years the weather is such that the starting of wild growth and the best time for planting such crops as sugar beets and hardy garden truck falls between the 5th and 20th of April in Northern Colorado. Every beet grower should try to get the bulk of his planting done between these dates. In case of a large acreage it is always best to plant a portion earlier than this and a portion later rather than have all that cannot be planted at the best time either to early or too late.

197

On the Sugar Company's Experiment farm at Longmont, small patches have been planted on similar soil with similar treatment since 1913. Three years the seed was planted between April 18th and 25th; three between May 2nd and 6th and 1 year on May 15th.

The yield of beets per acre was as follows for the different dates of planting:

Dates of Planting	Yield
May 15 (1 year)	14.59 tons
May 2, 3, 6 (Average 3 years)	17.00 tons
April 18, 20, 25 (Average 3 years)	18.19 tons

It will be noticed that the earliest planting gave the best yield and that the yield decreased as the date of planting advanced.

Several years ago experiments were conducted at the Experiment Station of Michigan to determine the best date for planting sugar beets. While the dates given in the report have no bearing on the proper time for planting in Colorado, the results are interesting in that they show the effect of too early and too late planting.

Beets planted from April 22 to 26 average 8.76 tons per acre; those planted from May 6 to 10 averaged 9.78 tons per acre; and those planted from May 27 to 31 averaged but 7.37 tons per acre.

Too early planting often results in loss of stand by frost. Black root is quite apt to attack beets planted too early also, since cool weather may check their growth and weaken them. This weakening makes the young plants less resistant to the attack of the disease.

The loss of stand by frost and black root usually reduces the yield even though it may not necessitate replanting.

Too late planting is apt to result in a poor germination of the seed and in an uneven stand. The young beets which come up during very hot, dry weather are apt to be injured by high soil temperatures. This results in the death of many young plants before their roots reach the cooler and moister soil several inches below the surface.

While the date of planting is very important from the standpoint of yield, beet seed should not be planted before the seed bed is properly prepared. No amount of pains taken in the care of beets later in the season will overcome the damage caused by planting in a poorly prepared seed bed. Better results will be obtained by planting a little late than by planting earlier, if the seed bed is not fit.

How Deep to Plant

This is a much discussed question. If we could look ahead and see just what the weather conditions were going to be, it might be possible to vary the depth of planting accordingly. As this is impossible, the safe policy is the only one to follow.

Regardless of conditions prevailing beet seed should be planted not less than 1 inch deep nor over 1½ inches deep.

If planted less than an inch and the weather turns suddenly hot and dry the soil becomes dried to a depth of an inch. This either prevents the germination of the seed or kills the young plants just emerging from the seed.

In case we plant less than an inch and the soil becomes dry before the seed has germinated light showers frequently dampen the soil about the seed which then starts to germinate, but before the young plant is rooted the soil again dries and the plant dies.

Frequently growers plant deep when the seed bed is dry in order to put the seed into moist soil. If this requires planting over an inch and a half many plants never reach the surface and many of those that do are weakened by the time the surface is reached, so that they fall an easy prey to disease.

If the seed bed is too dry to germinate seed planted at the proper depth, either irrigate before planting or plant properly and wait for rain to germinate the seed.

SPRING POETRY

A near-poet in New York, who wishes to remain "strictly anonymous," sends us the following. We print it gladly, but without any indorsement of the addition of lime to acid phosphate. We trust some of our Southern friends will regard this as poetic license:

> Little drops of acid,
>   Little grains of rock,
> Make the acid phosphate
>   To put upon your piot.
>
> Add a little potash,
>   A pinch or two of lime;
> Just a dash of nitrate,
>   And then just give it time.
>
> Plow your furrow deeply,
>   Cultivate it well;
> Hard and earnest effort,
>   In the end will tell.
>
> When the harvest's gathered,
>   And you count your gain,
> Don't forget to thank the Lord,
>   Who sent the sun and rain.

—The American Fertilizer.

WHY
The
VICTORY
LIBERTY
LOAN
?

**IS AMERICA LOYAL TO HER SOLDIERS
ONLY WHEN THE NATION IS IN
IMMEDIATE DANGER?**

The Job Isn't Finished Until
the Bills are Paid!

Roads

J. F. Jarrell

HEN we think of taking a trip to some point at a distance, the first thing we consider is the road over which we are going to travel We almost invariably route ourselves over what we think is the best road, regardless of whether it is a railroad or an auto road. Since the auto road is becoming, in a local sense, of more importance to a farming section than a railroad, it is the former about which we, as farmers, are the most concerned.

In the past our country roads have been a product of local capital and local talent. In some cases the capital was available and the talent was not. In other cases, the talent was available and the capital was needed. Yet in other cases both the capital and the talent were lacking. In still other cases the proper combination of capital and talent have been available and good roads have been the result. Government aid, State aid and corporation aid have greatly benefitted the roads in some sections.

When we see an exceptionally good stretch of road we "step on it" (the throttle) and feel greatly relieved to be able to cast a glance to either side of the road and enjoy the scenes as they float past us. What a contrast to the other condition of keeping one's eyes strained continuously on the space ahead which has been at one time named a road, but which has long since been divorced from that title. Automobile drivers are prone to cuss the road; the County Commissioners, the County, the State, and, in fact, every one except himself who, to my mind, is the one who is most to blame. The writer does not wish to unduly blame the autoist for poor roads, for he fully realizes what a big problem the roads of our country present. There are many things to consider when we think good roads. The capital, the building and the maintenance are all required, and the whole thing reverts to the first and main thing—viz.: capital.

Does a flivver pay anything like what it should as a fee for the damage it does to our roads? How long a piece of good road would two and one-half dollars make? Does it seem reasonable? A stingy little fee of $2.50, $5.00 or $10.00, to be applied to our road fund to offset the wear and tear which we do for 365 days.

A Northern Colorado editor recently remarked that 'the stupendous sum of $340,000 had been paid into the Colorado road fund during 1918." Think for a moment what that means. Enough money to build at least 15 miles of fairly good road without allowing very much for maintenance. Fifteen miles of good road! Just imagine, if

201

you can, what that means. Maybe enough room for a garage floor big enough to store one-half of the automobiles in the state. Isn't that immense? It is as a garage floor, but as a road, it is nil.

From time immemorial people have experimented at road building and the experiment is still being made just as though no one else had done anything along that line. In our haste we seem to forget the important things in road building and spend our time and money for temporary, make-believe roads, which are a disgrace to our assumed intelligence. The writer does not wish to knock the roads of Northern Colorado, nor of any one particular section of the country. But I think everyone will agree that the best roads we know might be better, and if they were, we would need fewer tires, less gas, fewer shock-absorbers, and less maintenance expense on our cars.

Let us begin to think soundly along lines of fairness to the roads and, if necessary, pay an auto tax which will more nearly pay for the damage we do to the roads and a long step toward the securing of good roads will have been made.

A Warning

A. C. Maxson

Last summer an army of webworms appeared in the beet-growing sections of Colorado, Montana, Nebraska and Wyoming. Lack of preparation and shortage in equipment and poison resulted in a wide spread and heavy damage to the beet crop over several thousand acres.

The first wave (brood) of this army came over the top in June, 1918, and the second in July and early August. Many of the last wave were destroyed. However, a large remnant of the enemy has spent the winter entrenched within their cocoons. The worms which spent the winter in our fields will be destroyed when the plowing is done. However, thousands entrenched themselves among the weeds on waste land, roadsides and ditchbanks. All such land should be plowed, disked or harrowed. This will either expose, bury or destroy the cocoons. If this is done the number of moths appearing during May and early June will be greatly reduced. This will mean reduced chances that our fields will be damaged again this season.

Keep all weeds down about fields and keep the beet crop clean. This will not only help prevent webworm injury, but reduce the number of other injurious insects about your fields.

Increase in Crop Acreages During The War

(Figures from "The Monthly Crop Reporter")

The acreage of all crops in the United States was:

1899	283,220,000	Acres
1909	311,290,000	"
1916	344,790,000	"
1917	356,340,000	"
1918	367,738,000	"

About 56 million acres were added to the cropped acreage from 1909 to 1918, an increase of about 18%.

The gross value of all crops was:

1899	about	3.0	billion dollars
1909	"	5.5	" "
1916	"	9.1	" "
1917	"	14.2	" "
1918	"	13.5	" "

The gross value of the crops increased about 8 billion dollars from 1909 to 1918, or about 146%.

The cropped acreage in Colorado increased

from............................2,614,000 acres in 1909
to ....4,060,000 " " 1918 or about 55%

In Montana it increased

from............................1,848,000 acres in 1909
to ....4,845,000 " " 1918 or about 161%

In Wyoming it increased

from...................... 786,000 acres in 1909
to ....1,569,000 " " 1918 or about 100%

The cereals in Colorado (corn, wheat, barley, oats, kaffircorn) increased

from............................1,057,000 acres in 1909
to ....1,762,500 " " 1918 or about 76%

The largest increase in the mountain states must, of course, have taken place on the so-called dry farms.

In spite of the labor shortage more acres were handled than ever before. Undoubtedly the tractor was one factor making this possible.

If production has increased to the same extent as the acreage, and if this production is maintained, the marketing problem will be more difficult and important than ever.

Crops with a guaranteed market and price therefore should be increasingly attractive.

By courtesy of N. Y. Tribune

The Fundamentally Best in Our American Life

"I WARN my countrymen that the great recent progress made in city life is not a full measure of our civilization; for our civilization rests at bottom on the wholesomeness, the attractiveness, and the completeness, as well as the prosperity, of life in the country.

"The men and women on the farms stand for what is fundamentally best and most needed in our American life. Upon the development of country life rests ultimately our ability, by methods of farming requiring the highest intelligence, to continue to feed and clothe the hungry nations; to supply the city with fresh blood, clean bodies and clear brains that can endure the terrific strain of modern life; we need the development of men in the open country, who will be in the future, as in the past, the stay and strength of the nation in time of war, and its guiding and controling spirit in time of peace."

The White House, February 9, 1909.   THEODORE ROOSEVELT

The closing paragraphs of the Special Message transmitting the report of the Commission on Country Life to the Senate and House of Representatives.

Home Grounds

W. S. Van Doren

N the February issue, H. H. Griffin contributed "Beautifying Country Roads." This thought may well be carried further and applied to the home grounds.

The writer wishes to suggest a plan of such a scheme to be carried out in April because the plan is the first essential in any work and April is the best month for the transplanting necessary in carrying out this plan.

If the home grounds are not large, a simple unbroken grass plot of blue grass and white clover with a few flowers grouped in the corners or around the porch is the most attractive. The lawn should be well leveled to provide for its proper care.

If the grounds are more pretentious, the monotony may be broken by a specimen tree or shrub on the lawn here and there, or a few shade trees around the house. Low-growing shrubs may be grouped occasionally about the house, but the writer does not believe in planting too close together nor in selecting too many varieties for the same space. Strive to create beauty, simplicity and dignity for these outward qualities involuntarily bespeak the character of the inmates of the home.

The walks and drives are usually the most direct line to their objective, but in large grounds or parks it is appropriate to have them curve. The curves may be hidden at intervals by masses of foliage.

As gardeners, we are not very venturesome, and we respond slowly to new ideas or cling to those names which ring familiarly in our ears. Space will not permit us to give more than a name or two in suggesting each group. The specimen tree may be represented by the cut-leaved weeping birch, the mountain ash with its heavy clusters or red berries in the fall or the Colorado blue spruce, most handsome of them all. On the outskirts of the lawn or by the fence, plant such as Swaddler's Norway maple, black and honey locust, walnut, Russian olive or the stately elm. For hiding the roads or massing about the house, the low-growing shrubs that give the best effect are the dogwood or the Japanese barberry (Thunbergi). In this connection, avoid planting the common barberry (Vulgaris Purpurea) because the Council of Defense has emphasized the danger of its relation to the black or stem rust of wheat. The shrubs that bloom to be grouped with these are almond, spirea, syringea, moc-orange (Philadelphus Coronarious), hydrangea

As Early as You Can—

Make up your mind what you are going to do.

Do it.

Have harness, plows, levelers, drills, in perfect working order before the spring opens.

Get your horses fed up so they can stand the work.

Spread whatever manure you have.

Roll cloddy fields.

Double disk before plowing.

Harrow at least twice, what you plow each day.

Most farmers plan to raise a bigger beet acreage this year than they have had for a long time.

This means the harvest will start earlier than during the last two years.

Therefore—

You Better Plant as Early as is Advisable in Your Locality

(P. Grandiflora), the bush honeysuckle (Lonicera Morrowi), or the butterfly bush (Buddleia), which, with its purple spikes, blooms from July till killed by the late frosts.

For the low porch, such vines as the scarlet trumpet honeysuckle, the Virginia creeper, clematis (Paniculata) with its white blossoms, and the jackmani with its rich purple blossoms. This last variety must have a position where it will be free from the north exposure and the dripping of the eves. For covering the high porch or walls, use the Boston ivy (Ampelopsis Veitchi) and the broad-leaved Dutchman's Pipe (Aristolochea Sipho). For dividing lines instead of fences, use spiraea, hydrangea, Japanese barberry, golden elder (Sambucus Nigra Aurea). Use hop vines or wild cucumbers for covering unsightly buildings, etc.

Flowers, both annual and perennial are used for filling in, and while no garden is really complete without annuals; such as a diagonal bed of mixed phlox and sweet alyssum or a circular bed of asters and lobelia, or geraniums and ageratum; still more emphasis should be placed on the perennials. The idea throughout this article has been along the line of hardy plants and of permanency; therefore, a few hardy perennials will be outlined. The tall hollyhock and larkspur will hide the high-board fences and fill in between wide openings. Plant in front of these the lower-growing canterberry bells, perennial phlox—in various colors—the iris, shasta daisy; and in front of these such as for-get-me-nots and pansies. The lupines look well bordered along stone walls. The bright colored Oriental poppies combine well with spirea and syringea. Peonies bloom in June and give the most striking effect when massed in their separate colors, either red, white or pink. The foxglove is another one blooming about this time that is effective in massed planting. In nearly every case, the varieties given will come up every year with the simple care of weeding and enriching the ground, and will give the most pleasure with the least work. They will also furnish cut flowers for the house, but for this purpose it would be a sad mistake to leave out the graceful and dainty perennial baby breath (Gypsophelia).

There is a great opportunity for individuality in these matters, and with ordinary good taste, combined with a little experience, one may work out very pleasing effects.

By beautifying his grounds, the owner cannot fail to derive financial benefits in every case, besides the pleasure he and his family will receive from such surroundings. To cite only one case, to emphasize this point, the writer knows of a man who spent his time and one hundred dollars beautifying the home grounds. The result was that he sold the place for one thousand dollars increase. Can you beat it?

Why Should Potato Graders be Adopted in Colorado?

Recognized Standards Are Essential in Successful Marketing—

There are many factors pointing to the desirability of adopting potato grades and standards for use in Colorado. In the first place a recognized standard is one of the essentials for the successful marketing of any product. The consuming public of this country remembers those things which satisfy. On this basis they purchase again and again products under a "label," "trade name" or "trade mark."

Standard Grades Provide a Guarantee of Quality—

Standard grades provide a guarantee of quality. Under a uniform system of grading and inspection, shipments cannot be rejected upon mere pretenses. The inspection certificate is a guarantee to the grower and consumer of the quality of the potatoes sold. Without grades, the buyer may reject potatoes on account of quality, and the grower has no redress.

How the U. S. Potato Grades Were Derived—

The present grades of potatoes known commercially as U. S. Grade No. 1 and U. S. Grade No. 2, are the result of three and one-half years' investigation in all of the principal potato producing sections of this country. They represent not a grade which applies to one locality or another, or to one particular market, but were promulgated by the Bureau of Markets of the U. S. Department of Agriculture and the U. S. Food Administration to meet the marketing requirements of potatoes for the entire country. These grades were in operation during the marketing of the larger part of the 1917 potato crop shipped during the fall of 1917, and the spring of 1918. They have been applied thus far in the marketing of the 1918 crop. They have proved satisfactory in practically all of the principal shipping sections of this country.

Sorting Costs, But Brings Results—

The advantage of sorting culls and low-grade potatoes from the stock to be marketed has been recognized by the potato grower. Its value is easily shown by an inspection of the stock of any potato dealer. Potatoes grading U. S. No. 1 invariably command from 25 cents to 50 cents premium over potatoes grading U. S. No. 2, while field-run potatoes are for the most part a drug upon the market and practically valueless except in cases where the demand for potatoes is unusually strong. In fact, 75 per cent of the potatoes

208

sold as field run are sorted before they reach the consumer. The labor and time necessary to cull out the worthless stock is time well spent because of the premium which well-graded stock will bring over ungraded stock. The grower should grade and get this higher price.

The Future of the Colorado Potato Industry Depends on the Adoption of Grading Laws—

Potatoes are Colorado's greatest perishable crop. Over 12,400 cars of potatoes were shipped from Colorado from the 1917 crop. Taking an average value of $400 per car (figuring on the basis of $1.20 per cwt.), the 12,400 meant over five million dollars in revenue from the 1917 crop to the potato producers of Colorado. The markets in the states east and south of Colorado are logical consuming centers for this grade shipment of potatoes. The satisfied consumer on these markets is the man that fixes the future of this great industry in Colorado. Idaho, Utah, Wyoming and Nebraska are shipping potatoes to these markets. If the buying consumers prefer Idaho potatoes to those produced in Colorado, the potato production of those states will increase and the Colorado industry will decline.

Successful marketing methods have taught us that a product of a standard uniform quality is one of the keys that keeps the consumer satisfied. The consumer wants potatoes of good quality. If he is satisfied in his purchases, he would like to know where and how he can get the same satisfactory product again. This can only be done where recognized grades and standards are applied. The future of the Colorado potato industry hinges upon the adoption of recognized grades in marketing.

—STEWART L. SWEET, U. S. Department of Agriculture.

THE FUTURE OF WHEAT

The outlook of the wheat farmer after June of 1920 is not at all bright. In all probability at that date there will be a vast wheat reserve on hand in North America, South America and Australia. Unless this reserve is burned or otherwise destroyed, we can not foresee any way of handling it which will not have a decided effect on lowering prices sooner or later. And it will not be burned.

The wheat farmer unquestionably faces a more precarious future than the corn farmer, the cotton farmer or the live stock farmer. The corn farmer must realize, however, that anything which affects the price of wheat, indirectly affects the price of corn. And anything which affects the price of corn must sooner or later affect live stock prices.

The entire wheat situation is fraught with the most interesting possibilities, not only to the wheat farmer, but to the man who grows chiefly corn and oats. And, incidentally, the live stock man is going to be touched.—Wallace's Farmer.

Early Planting of Beets Very Profitable

Mark Austin

I N an issue of the Utah Farmer under date February 15, 1919, we called attention to the benefits of early and proper preparation of the soil, early planting and early and careful attention to the crop. A large percentage of the beet growers are aware of the beneficial effects of the early work, and to substantiate our claims for the benefit of our beet growers, we submit below some figures on April and May plantings. It shows the results obtained in our various factory districts of beets plants in April and those planted in May. About 100 names from each factory district for each of the aforementioned months were used in our comparison which we think gives a very fair average:

April Planting

Average Tonnage	Average Sugar	Average Purity
13.05	16.02	87.16

May Planting

Average Tonnage	Average Sugar	Average Purity
10.81	15.48	86.70

You will notice that the April, or early plantings, not only yield on the average about three tons more per acre, but they contain a higher sugar content than the May planted beets, notwithstanding the fact that the beets planted in April were harvested two or three and in many cases four weeks earlier than those planted in May. Had the early planted beets been left in the ground until the time of harvesting the late beets, no doubt there would have been larger increases in tonnage and sugar content.

The figures above given were taken from all factories where we operate and the averages compiled. We thought the farmers would be glad to have this valuable information printed. If this is carefully observed, and the suggestions put into practice, it would be very profitable to the farmer and the industry, in which we are all interested. The early planted beets usually ripen first, which enables the farmer to get his harvesting started sooner than if he planted late.

The beet growers should be very careful in the preparation of their soil, to see that it is worked down to a seed bed before the soil dries out, where fall plowed or spring plowed. It is a good plan to level and harrow the land before plowing, where it is to be spring plowed, which will keep it from crusting. If the soil is handled in this manner a dust mulch will be formed on top which will put it in good condition for plowing.—From "The Utah Farmer."

Formaldehyde vs. Bluestone in Seed Treatment

"In certain communities of the state there seems to be some question as to the advisability of using formaldehyde in preference to bluestone in the treatment of small grains to prevent smut," says James Godkin, of the Colorado Agricultural College. "Before continuing this discussion further, we would like to have it understood that it is not the intention of the State Agricultural College or the U. S. Department of Agriculture to create a prejudice against bluestone in favor of formaldehyde in our smut control work now well under way in Colorado. However, we do wish to emphasize some points which we now know to be facts. Formaldehyde has several distinct advantages which would seem to offer sufficient proof to warrant its use in preference to bluestone.

"1. It is simpler in its application, because it does not need to be dissolved, and the treatment does not need to be followed by an application of lime water to the treated seed to prevent seed injury.

"2. It is not nearly so apt to injure the grain. On the other hand, bluestone causes injury to the germination and development of seedlings of barley, oats, and wheat.

"3. Any grain treated with formaldehyde which happens to be left over after the treatment may be safely fed to stock after the grain has dried. Grain treated with bluestone is poisonous to stock.

"4. Formaldehyde is cheaper than bluestone. It is true that formaldehyde costs more per pound than bluestone, but 1 pound of formaldehyde is equivalent to about 8 pounds of bluestone in the treatment of the same amount of grain.

Amounts of formaldehyde to use:

16 oz. (1 pint.)	Sufficient for	40 bushels
8 "	" "	20 "
4 "	" "	10 "
2 "	" "	5 "

"Whether you favor formaldehyde or bluestone be sure you treat and use recommended methods in your treatment.

"Treat every season—it pays."

Stem Rust Does More Damage Than Enemy Incendiaries

Secret enemies destroyed many elevators of wheat during the war, but it is the enemies that escape detection from whom we have most to fear. Such an enemy is stem rust, but now that its camouflaged headquarters on the barberry plant have been discovered by the penetrating eye of science, it will be shown no more mercy than treacherous enemies deserve. No longer will this destructive parasite on the under surface of an innocent looking barberry leaf produce its thousands of rust spores which carry disaster to the grainfields and hunger to the home. Stem rust, like other parasites, is helpless without its accustomed food. Though it is too small to be attacked directly, it can be starved by the removal of the barberry bush. The barberry must go. It harbors the rust parasite.

—E. L. SARGENT,
Colorado Agricultural College, Fort Collins, Colorado.

Remove Smut Balls Before Treating Seed With Formaldehyde

"One of the most common smuts in Colorado is bunt or 'stinking smut' of wheat," says Dr. W. W. Robbins, of the Colorado Agricultural College. "It has the habit of forming so-called smut balls." These are rather solid masses of smut which may be entirely covered with the grain coat and thus escape attention. In treating wheat affected with bunt with formaldehyde solution, the fumes of formaldehyde may not penetrate to the center of the smut ball. As a consequence, some of the smut balls may carry viable spores into the ground, although the seed was treated. Consequently, pains must be taken to remove all smut balls before treating with formaldehyde solution. Their removal may be accomplished by thorough fanning or by floating the grain out in water or formaldehyde solution and skimming off the smut balls.

p. 373-374 clipped

THROUGH the LEAVES

AUGUST, 1919

THE GREAT WESTERN SUGAR COMPANY

THROUGH
THE
LEAVES

AUGUST, 1919

Published Monthly By

THE GREAT WESTERN SUGAR CO.

DAILY CALL PRESS
LONGMONT, COLORADO
1919

TABLE OF CONTENTS

Editors' Page

Why not spend some time dragging the roads and getting them in shape for the heavy hauling this fall. The roads can be put in good condition after the rains we are having.

* * *

In making plans for next year remember that fall plowing is the first step toward a profitable crop.

* * *

At this time some thought should be devoted to the selection of wheat for seeding this fall. Special attention should be given to get wheat of one variety that is free from foreign seeds and smut.

* * *

There are many farmers expecting to silo their beet tops this year. The silos should be dug before harvest time so that the tops can be siloed while fresh.

* * *

Beet top silage will have an added value this year when we expect hay to sell at very high prices.

* * *

It is only a short time until beet harvest. Wagons and harness undoubtedly need repair after the summer's work before starting the beet harvest. Why not get this done now and be ready for the fall work.

* * *

Be sure to disc your ground before plowing as it prevents clods, which will be difficult to work into a good seed bed next spring, if we have a dry winter, from being turned under.

* * *

Due to much shifting of labor this year some growers may find it difficult to get just the help they will need at harvest time. If any growers need additional help, or if their beet help has left them, they should get in touch with their fieldman now. Do not wait until harvest time to attend to this matter.

* * *

There is sufficient moisture in the ground to enable the farmers to fall plow their ground.

* * *

Some farmers will be unable to plow their land until the grain is threshed. In this case it would be well to disc between the shocks and thus conserve some moisture.

"The Botany of Crop Plants"

W. W. Robbins

Professor of Botany, Colorado Agricultural College
(Reviewed by H. Mendelson)

 SUCCESSFUL animal breeder is always familiar with every visible part of the anatomy of the animals he produces. He soon sees the good and bad points of each individual. One of his main enjoyments of life is to discuss the merits of his animals with anybody competent and interested enough to listen.

The writer has never met a successful breeder who could not grow eloquent on his subject, no matter how silent he might be on anything else.

Crop raising farmers seldom have the time or take the time to become so familiar with the plants they raise, that they get some joy out of particularly good specimens.

Many sugar beets presented at county fairs can be well used to demonstrate what a sugar beet should not look like.

Lack of knowledge how the crop grows often results in faulty farm practice. For instance, for many years it was believed that

---

Plowing can be done more cheaply now than in the spring

---

early irrigation of beets produced a shallow development of the root system, while as a matter of fact a beet only thirty days old has roots longer than twelve inches.

We listened at one time to a discussion between two men who had been raising beets for at least ten years; whether the small rootlets of the beets grow at a right angle to the row or in the row. Of course both were wrong.

Familiarity with the growing plant acquired by close observation is a great help in making successful crop raisers.

Professor Robbins' book on "Botany of Crop Plants" gives a great deal of interesting information on the structure and life of plants usually grown on farms.

The first part of the book gives some general information applicable to all plants grown from seed in eight chapters covering the subject of the seed plant body, the internal structure of the plant body, roots, stems, leaves, flowers, fruit, seed and seedlings and the classification of plants.

The second part discusses in thirty-two chapters details of the various crop plants as follows:

The grass family (to which all grains, including corn, belong):

366

Wheat, oats, barley, rye, corn, sorghum, rice, millet, timothy, sugar cane.

The lily family, including onions and asparagus.

Mulberry family, including mulberry, hops, figs and hemp.

Buckwheat family, including buckwheat and rhubarb.

Goosefoot family, including spinach and beets.

Gooseberry family, including gooseberries and currants.

Mustard family, including all cabbages, cauliflower, Kohlrabi, turnips, rutabagas, rape, mustard, radishes, horseradish and water cress.

Rose family, including raspberries, blackberries and strawberries.

Apple family, including apples and pears.

Plum family, including plums, prunes, cherries, apricots and peaches.

Pea family, including peas, beans, vetches, clover, alfalfa, white clover, soy beans, cowpeas and peanuts.

Flax family.

Rue family, including lemons, oranges and grapefruit.

Grape family.

Mallow family, including cotton and okra.

Carrot family, including carrots, parsnips, and celery.

Huckleberry family.

Olive family.

Morning glory family, including sweet potatoes.

Night shade family, including potatoes, tomatoes, pepper and tobacco.

Gourd family, including cucumbers and melons.

Thistle family, including lettuce, chicory, oyster plant and artichoke.

The chapters on most individual crop plants are subdivided. For instance, the one on wheat discusses the following:

Habits of plants.

Roots.

Stems.

Leaves.

Inflorescence.

Spikelet.

Flower.

Opening of flower and pollination.

Artificial cross-pollination.

Fertilization and maturing of grain.

Ripening stages.

The mature grain.

Relative proportions of the parts of the grain.

"Hard" and "soft" wheats.

Milling of wheat.

Kinds of flour.
Germination of wheat.
Classification of the types of wheat.
Origin of wheat.
Environmental relations.
Uses of wheat.
Production of wheat.

Ample illustrations accompany each chapter. Also a list of references or of the literature quoted for those inclined to further study.

This book of course does not discuss farm practice or soil preparation, although occasionally points of practical importance are mentioned. For instance, in discussing the mature grain of wheat, Professor Robbins says:

"Although the results are conflicting, there are insufficient positive results to warrant the belief that large plump seeds will give uniformly greater yields than small seeds, especially when such seeds are secured by means of the ordinary fanning mill. It is known that not all the grains in a spikelet are the same size and weight—the second is the heaviest, the first and third about equal in weight, and the fourth and fifth, if present, are lightest of all. It is obvious that all grains from a spikelet, regardless of their size, have the same heredity. And a light seed from a spikelet usually will, under similar

---

Double plow your alfalfa ground this fall

---

environmental conditions, develop into a plant with as much vigor as one from a heavy seed from the same spikelet. In the selection of seed wheat, the individual plant should be the basis of selection, when such method is practicable, rather than to depend upon seed from the bin or sack, which is the offspring of many different parent plants."

This, of course, does not mean that fanning seed grain for the purpose of removing weed seed or smut balls is not a good plan.

In some places some information of practical value might have been given without hurting the scientificial dignity of the book, as for instance, about the difference between alfalfa and sweet clover seed.

The book is written as text-book for agricultural and other schools and was not intended as popular reading. All botanists seem to enjoy a frequent use of technical terms derived from Latin and Greek words, much more than is necessary in the reviewer's opinion.

However, a glossary is added to this book, so if one really desires the valuable information in the book, it is not difficult to understand it.

The teaching of botany in city schools, illustrated by crop plants rather than pretty flowers, is very desirable. The clash of interest between the city consumer and the country producer is going to be very intense in the near future. An acquaintance with what is growing on the farm may give the city man a little more sympathy with the justified demands of the farmers for wages, a little more on a level with city workmen than sometimes in the past.

Also the country school can do a great deal more in interesting the children in what is going on, on the farm.

The book is heartily recommended for our libraries and schools and to farmers willing to get a little better acquainted with those living beings from which they make a living—the crop plant.

Disc between your grain shocks

Weeds Our Everlasting Enemy
H. T. Delp

The boldest robber and the most savage outlaw that ever came to farmers is old man Weed. His tribes come in at all times and stay forever, robbing our store of plant food and soil moisture. We have had weeds since Eve ate that proverbial apple, and I suppose they will be here until the world is swallowed up in flames. Therefore, it behooves us all to roll up our sleeves, get our fighting togs on and destroy weeds.

Weeds are more prolific than guinea pigs and rabbits combined, so it is safe to say that "to kill a weed in time saves 999." Manure and irrigation water, the two most essential things we give to our crops each season, are heavily infested with weed seed. So we are sure that our crops will have plenty of severe competition for food and water every year. The more we do to help them along in their battle for supremacy, the better they will yield for us at the end of the season.

Cultivation is the most effective weapon the farmer has with which to fight weeds. It has been shown time and again that the chief function of cultivation is to destroy weeds. Rotation of crops and clean seed both help. Also to clean up the breeding places, as fence corners and waste patches helps to keep weeds out of our fields. Weeds are faithful partners of insects, and the more weeds we have the more insects we should expect. Then let us swat the weed and have bigger and better crops. Whisper the same thing to your neighbor, for he is your side-kicker every time when it comes to handling the weed crop.

369

Is the Sugar Beet More Apt to be Injured by Insects Than Other Crops?

A. C. Maxson

THE sugar beet growers of Colorado, Nebraska, Montana and Wyoming have just finished a fight against the greatest webworm outbreak in the history of the beet sugar industry in America.

In those territories furnishing beets for the Great Western Sugar Co. 166,728 acres were infested by webworms during the latter part of June and fore part of July.

The campaign against this army of worms began June 15th and continued for from 20 to 25 days.

Owing to a shortage of sprayers and poison only about 100,000 acres were sprayed. Of the 166,728 acres infested, spraying prevented damage on 76,764 acres; the damage was light on 47,811 acres, moderate on 25,997 acres. The damage was heavy on about 15,898 acres. This last was due to late spraying or lack of spraying in all cases.

In the fight against the countless numbers of worms which overran the beet fields of the entire Rocky Mountain section, the Great Western Sugar Co. supplied the growers with over 600,000 pounds of Paris green and the use of about 400 sprayers.

After having watched the leaves of one's beet crop disappear as if by magic; after having seen the webworms migrate in armies from adjoining fields and weed patches into one's own beets; after having worked night and day to save the crop; and after having spent $2.00 an acre for Paris green and team and man labor also, one is apt to come to the conclusion that sugar beets are more apt to be injured by insects and diseases than the other crops grown in their locality. Lest we become convinced that this is the case let us take note of the pests and diseases' attacking other crops in our section of the country and others.

Wheat

Most of us remember the outbreak of cutworms which caused the total loss of hundreds of acres of wheat and other small grain during the years 1911, 1912 and 1913.

The spring of 1918 many acres of wheat were damaged even to the point of being a total loss by the stem maggot. The spring wheat aphis or green bug injured a considerable acreage also. Every year a considerable, although unnoticed damage is caused by such insects as the Wheat Straw worm and the Wheat Joint worm.

Loose smut and stinking smut (bunt) cause heavy annual losses in spite of scattered attempts to prevent it by treating seed.

Wheat rusts often make late wheat unfit for market. This has occurred in Northern Colorado during the last five years.

Alfalfa

The first cutting of alfalfa has been almost a total loss several times during the past few years because of the stem blight.

On the Western Slope the alfalfa weevil has made its appearance. This insect can be successfully controlled only by sacrificing the first cutting of hay.

The alfalfa looper has seriously damaged alfalfa on the Western Slope and parts of Montana. Locally, a small caterpillar has caused the loss of many tons of first cutting hay in Northern Colorado. Grasshoppers have killed many acres of newly seeded alfalfa the first season. Cutworms also injure the first cutting severely.

Corn

If grown several years on the same land corn suffers from the corn root worm in Colorado. The ear worm damages the grain and spoils sweet corn for the market.

Corn smut reduces the value of corn for silage and reduces the yield of grain as well.

In parts of the corn growing states the corn root aphis, wire worms and white grubs cause heavy annual losses.

Conserve moisture by discing before plowing

Potatoes

For several years the potato crop of Northern Colorado was a failure. In order to learn the cause of these failures the government built a laboratory near Greeley where the diseases which were found to be the real cause were studied. Both early and late blight injure potatoes frequently in Northern Colorado. This year many acres of potatoes were damaged by the potato beetle.

Hundreds of dollars are spent in fighting the potato flea-beetle, potato aphis, leaf hoppers and other pests attacking this crop in the eastern portion of the country.

Some potato growing sections cannot sell their crop for seed because it is infested with the tuber moth or some one of the many tuber rots.

Beans

The bean beetle causes heavy losses to bean growers in Northern Colorado. No less than four diseases; bacterial blight, pod spot, root and stem rot and rust attack this crop and not infrequently result in complete crop failure.

871

Peas

Not infrequently the pea aphis damages the pea crop quite seriously. In certain sections of the country pea raising for seed has been abandoned because of the pea weevil.

Cabbage

Cabbage growers loose heavily from the ravage of the cabbage worm, cabbage aphis and diamond backed moth. Root maggots and several diseases help to increase the annual losses.

Tomatoes

Tomatoes suffer from flea beetles and the disease called wilt as well as diseases of the fruit.

Apples

The greater part of the apples of Northern Colorado are made unfit for market by the codling moth. The trees are injured by the apple aphis and leaf roller.

We might go on at length naming such losses as are occasioned by the raspberry stem blight, the currant sawfly, the pear slug on cherries, the root worms in radishes, squash bugs on pumpkins and squashes, the cucumber aphis and cucumber beetle on cucumbers and melons, the aphids on elm trees and snowball bushes until we had named every field, orchard or garden plant and most ornamental plants as well.

With this formidable array of crop pests and disease before us the occasional losses caused by webworms, cutworms, alkali beetles, flea beetles, black root and other insects and diseases of the sugar beet do not look out of proportion.

While in the case of nearly all other crops the grower is left to his own resources in combating the pests which threaten to reduce his year's profits, the Sugar Company gives every assistance possible both financial and otherwise to its growers.

Before we condemn the sugar beet because this, that or the other insect or disease attacks it, let us weigh the facts carefully and decide impartially.

If you have electric lights you know what it means to blow a fuse. A fuse is a piece of wire that burns out more quickly than the ordinary wire, and when some extra load is put onto the current, instead of the motor or some other more expensive part being destroyed, the fuse blows out. This idea can be applied to all farm machines, as most machines have what correspond to a fuse. By keeping a good supply of these weak parts on hand, machines are delayed very little because of breakdowns.

Fall Plowing Helps Labor Distribution

P. H. McMasters

ALL plowing should have its place in every farmer's system of farm management if the maximum results at the minimum cost are to be obtained.

The shortage of farm labor especially during the planting season necessitates the farmer employing different methods of management than when labor was more abundant. It behooves the farmer to give employment to his labor throughout the year if the best labor available is obtained. Very few men are satisfied to work on one farm a few months and then be compelled to look for another place. Steady employment throughout the year is one means of securing labor and being able to keep it.

Distributing the farming operations throughout the year can be accomplished only by doing as much work as possible during the slack seasons, that is usually done in the rush season. In other words, do away with the rush seasons.

Fall plowing lends itself exceedingly well to this system of management in that it permits the farmer to get his land plowed at a time when there is not a great deal of other work to be done. Many farmers realize the adavntage of fall plowing but complain that they are unable to get a large amount of it done until the grain is threshed. Last year many farmers did not get their grain threshed until late September, and it looks at the present time as if the same would be true this year. Most farmers finished cutting their grain July 20th,

ing season.

Summarizing, we find that stacking the grain would permit a great deal of fall plowing which in turn would lessen the amount of work to be done in the spring. This would enable the farmer to get along with one man where it usually takes more.

One Reason for Fall Plowing

A. C. Maxson

Just at this time we are all very much interested in doing everything we can to prevent a recurrence of the webworm outbreak which kept farmers and factory people up night and day for nearly a month.

While the webworm is uppermost in our minds just now, yet, we are equally interested in keeping down grasshoppers, cutworms and other insects which damage our crops.

Fall plowing is one of the most effective means of fighting certain crop pests.

The cocoons of webworms turned up by the plow and exposed to the elements all winter rarely contain living worms in the spring. In case they are turned under deeply the moth will not succeed in reaching the surface next spring, even though the worm may survive the winter.

Grasshopper eggs buried 6 inches deep prevent nearly all young hoppers from reaching the surface in the spring. Those that are not buried but are exposed seldom hatch.

The pupal stage of many cutworms are destroyed if the soil is plowed before the moths emerge.

Fall plowing buries crop refuse under which or in which many insects spend the winter. These insects will not spend the winter in fields plowed and thus made bare of material to shelter them from the weather during the winter.

Save moisture with a surface mulch

Lovell Lambs Marketed in July

C. R. Hedke

Location, climate, feeds and fertilizer determined the nature of the livestock end on the Lovell factory farms. The distance and transportation we have to markets score heavily against fattening, while climate and natural feeds favor stock raising.

Soil fertilizer in the form of manure is needed here more than in any other section of the beet territory of the Company. Our normal rainfall of about five inches annually produces little natural vegetation, and this is of such a nature that the natural humus in the soil is strikingly less than in any of the other beet sections. Therefore, agriculturally, the Big Horn Basin has a natural handicap of many seasons which can only be lessened by the application of considerable manure. Livestock must be had early on the farms in order to furnish a maximum production from the soil in a reasonable number of years.

The turn-over on cattle is slow and the country's economical condition during the past few years invited quick returns, so sheep raising was undertaken along with some cattle raising on the factory farms.

In the summer of 1917 sheep raising was begun by us in a small way, with results so satisfactory that it was rapidly enlarged upon, so that in the spring just passed there were dropped 800 lambs from about that many ewes. We also have some 400 additional yearling ewes for next season's lambing.

Our first lambing crop was bred for March, but on account of the condition of the stock, most of the lambs came in April. These lambs made such a good growth on dried beet pulp, fed both to the ewes and the lambs, as early as they would eat it, that even earlier lambing was decided upon, with the purpose of selling these spring lambs in July of the same year. This season's lambing began in February and extended to about April 1st.

During the last week of June a clean-up of old ewes and cull yearling wethers was had and 32 spring lambs were included in the shipment. These lambs averaged 70 pounds at Billings and returned $3.06 per head more than the 74 yearling wethers which averaged 82 pounds and were included in the same shipment. One hundred and fifty more lambs might have been sorted, with nearly the same average weight at this time.

The pictures herewith enclosed show the yearling wethers and lambs just before loading, and the lambs can easily be pointed out in the picture and the relative size noted.

For the next several years it is planned to lamb early and sell

June 30—Shipment yearling wethers, average weight 82 lbs., and spring lambs, average weight 70 lbs.]

all wether lambs in July. This requires special building equipment, but we believe it is worth the investment necessary.

Wool returns, too, have been very satisfactory from early lambing, as the yearlings are shorn early and made the good average of 12 pounds this year.

It is planned to run a large number of sheep on our farms during the summer on both sweet clover and alfalfa pastures. In this manner the maximum amount of fertilizer will be obtained from the live stock, and the best use will be made from the production of some of the poor areas on the Company farms.

Use disc set straight to pack ground after plowing

The Greeley Wonder Muskmelon

The picture on this page shows the new Greeley Wonder Muskmelon. Mr. W. B. Fatum of near Greeley, who is a sugar beet grower, is raising these melons. This new melon has taken the markets by storm. It originated with a market gardener, Mr. E. J. Hafendorfer of near Greeley, who has for several years been getting fancy prices for them in the home market as well as shipping them to Denver and elsewhere. The melon is large, early, of high quality and is a good shipper. It weighs four to five pounds and has yielded four hundred crates to the acre. The melons are creating quite a furore in the country north of Denver.

Beet Labor at $20.53

C. V. Maddux

NINE THOUSAND beet workers were shipped into the various districts last spring—6,000 into Colorado alone.

For this purpose $185,065.91 was spent—$20.53 per worker.

Eighteen states contributed the labor—company agents canvassed six states and parts of eight other states, and engaged help from four by correspondence. Sufficient beet help was engaged to tend all of the large acreage planted. Drought has heavily reduced this acreage and likewise the tonnage prospects for the present acreage; and as a consequence the cost of this item per ton of beets to be harvested will be abnormally high.

For transportation alone $146,345.66 was spent. Eighteen special passenger trains and a much larger number of special cars on regular trains were chartered.

Food was provided while en route and at destination pending arrival of growers at cost of $12,678.38.

It cost $26,000.00 to locate sources of supply, solicit the labor and conduct the various shipments to destination. The centers which had formerly provided this labor could not meet the requirements this year, largely because of the unsettled industrial conditions. New and further removed sources of supply had to be sought out. In February and March agents were sent out to locate them, covering first the nearby states and then spreading out as necessity developed. By the first of May it was considered that an adequate supply had been located. But unexpected competition developed about that time. In Kansas City, for instance, an eastern Sugar Company rented a hotel at which free board and lodging were given to Mexican workers and their families until a sufficient number was assembled to make up a shipment. Other eastern companies were also soliciting actively, and Canadian mills were offering to pay transportation both ways if labor completed the season. In St. Louis and Chicago strenuous competition was met from other industries as well as Sugar Companies, while from El Paso trainloads of Southern Mexicans were shipped each week to California, Utah and Idaho.

It is significant that Sugar Companies from California to Ohio should all be competing for this Mexican labor, for most of the labor engaged in the centers named was Mexican. This indicates that Mexican labor is considered the best available, and that generally it is giving satisfactory results.

In many individual cases satisfactory results were not obtained this year in our districts, but they were the exception rather than the rule; and frequent investigation has developed the fact that the labor is not always to blame. Some growers have not done their part, usually because of lack of experience in handling this kind of labor.

The Company, too, has made mistakes, due to lack of experience; but plans are in process now to prevent making the same mistakes in soliciting next year. For example steps will be taken to prevent misrepresentation on the part of solicitors and sub-agents regarding the principal requirements of the labor contract, both as to work and as to pay and concessions to be allowed. In other respects also the Company is striving to learn from last season's mistakes.

I always calculate that one ton of siloed beet tops is worth a half ton of hay. Feeding twenty to thirty pounds of siloed beet tops a day to steers they will only eat one-half as much hay and this is quite a saving with alfalfa at $20 or more a ton. If we reduce our hay 50 per cent and increase the value of all our feed by a ration of siloed beet tops thus we can readily save a lot of money and fatten our cattle fully as well. We figure that an average acre of beet tops well taken care of by siloing when green is worth as much for feed to fatten cattle as an average acre of alfalfa. Beet top silage is worth one-half as much as alfalfa hay, and we get twice the tons of green tops as we get of cured alfalfa hay. This is a feed that will give good gains on cattle without grain and with a long feed such as range steers must have, and besides it will accumulate manure so as to keep our land in a high state of production, which is our chief end and aim.—A. Hoeman, in "Utah Farmer."

STACK STORAGE

Unless unforseen changes alter the situation there will be an enormous grain crop and a car shortage, as well as a grain elevator shortage. What then should the farmers do?

It is more than likely that the world will get nearly back to normal in another year and there will not be the urgent need for surplus grain production in this country. That there will be a natural letting up when the wheat guarantee is removed, can hardly be questioned. It would be an extravagance not warranted by the necessity of the situation if farmers were to erect grain storage room not needed in normal times. They can stack their grain this year which will afford storage without cost, the lessening of the threshing bill probably being sufficient to pay for stacking.

Well-stacked grain will keep a long time and can be threshed at any convenient time. This is one solution of the storage shortage without risk, because the price guarantee will hold until July 1, 1920.

Of course, it is always economy to provide sufficient storage capacity to meet the needs of normal seasons.

The Sugar Beet Industry

Edith M. Bowen

(Editors' Note: This story won the second prize offered by the
Great Western Sugar Co., Billings factory for the best essay on
"Sugar Beet Growing," written by the students of the Huntley Project
schools at Worden, Montana.

The class had made a special study of sugar beets for one month.

The story which won first prize was printed in the July number
of Through the Leaves.)

Historians say that sugar beets helped nourish the builders of the
pyramids. They were cultured many years ago in Europe. The first
sugar factory in the world was built in Germany in 1802. Before the
recent world war, that nation was one of the foremost powers of the
world, due to her sugar industry. In the year 1830, sugar beets were
introduced into the United States. Today, it is one of our leading
industries. In a report by the Committee of Agriculture in 1838, it
was said that no country in the world was better adapted to sugar
beet growing than the United States, whether we consider soil,
climate or people.

About ten years ago, sugar beet growing began on the Huntley
Project. With fine soil, good climate and an irrigation system, the
industry flourished. Most farmers have only a short distance to haul
their beets to the dump. Here they are loaded into beet cars and
shipped to the Billings factory.

The value of the beets depend upon the purity of the juice and
the amount of sugar content. This year, beets are sold to the factory
at the rate of ten dollars per ton.

Sandy loam and heavy soils are the best kinds to produce good
sugar beets. Beets that are grown in a sandy loam are larger than
those grown in a heavy soil, but the sugar content is not so great.

Climatic conditions are very important to the success of the beet.
We have the warm sunshiny days and cool nights. Our irrigation
system is one of the best to be found. Water can be gotten at any
time. As beets need much moisture, it is better to have water near
at hand, than to depend upon rain. The Northern Pacific and Burling-
ton railroads run through the Project. This makes transportation
convenient to all the farmers.

There are drawbacks to be contended with in sugar raising.
Diseases, such as black-rot, sometimes attack the beet, and rainy
seasons make it almost impossible for the farmer to harvest his crop.
However, these drawbacks are few, and during the last three or four
years, the seasons have been fine, the sugar beet crop good, and
the help plentiful to harvest the beets. This spring, there are about
two hundred Mexicans, besides other laborers on the project. The
farmers have gotten their seed into the ground in fine shape and a
good crop is expected.

It has been said that more sugar beets were hauled over the Worden Dump in 1913 than in any other place in the world.

The beet tops are very good feed for cattle and sheep. Beets raise the value of the land. Sugar is one of the most used producets in the world. Until the devastated nations in Europe can rebuild their country, the people will depend greatly upon the United States for the sugar supply. It is up to the Huntley Project to do its share.

Fitting the Collar

The horse with an illshaped shoulder is a detriment to a farmer. There are many such horses, and it is a cause of much worry to the owner who is continually trying various expedients, such as changing collars and pads, and using remedies to heal sores and collar boils.

After many years of working horses we have learned that every horse is a problem in himself so far as the collar, that most important part of the harness, is concerned. A collar is a collar to some men. If it is the size in length, the horse must wear it even if it touches the shoulder only at one or two points. No wonder there are so many horses and colts with ugly collar bruises.

The prevention of these blemishes lies in the first working year of the colt in the majority of cases, and the normal shaped shoulder can be accustomed to the friction of the collar if the collar fits as it should. There are all sorts of necks and shoulders. There are the prominent high shoulder blades which are the hardest problem a farmer has to meet. But every shoulder, whatever its shape, can be fitted if the farmer cares to take the trouble to do so. If the neck is narrow, commonly called ewe neck, a wide collar at the top can be used if it is soaked in water for a few hours before fitting. After it is pliable, buckle it on the horse and fit the hames so they draw the collar just as it should fit for comfortable working. Let the harness remain till the collar retains the shape of the shoulder and the main source of trouble is avoided.

We have found it much better to prevent these bad shoulders rather than cure them, which is generally not an easy thing to do. We never use any grease or anything to soften the shoulder if a sore or bruise must be treated. We find slacked lime on a raw place as good as anything to heal and harden, and it has the merit of being cheap. There are proprietary remedies on the market that are good. Sulphur is often used to heal and harden a shoulder bruise and some horsemen use it exclusively for this purpose.

Years ago we eschewed pads of all sorts, unless under the most hopeless circumstances, and the case is bad indeed, which cannot be handled better without the pads, the pad being mostly an excuse to fill too large a collar. A perfectly fitting collar is half the battle.—Successful Farming.

Many Big Road Projects Planned by State Board

The following article is of interest to beet growers in Colorado, as in District No. 1 many of our beet territory roads will be affected.

 WORKING PLAN for highway improvements in Colorado that contemplates the expenditure of $2,342,500 for 1919 road projects has just been approved by the State Highway commission. The amount involved is by far the largest ever at the disposal of the authorities in any one year and represents the combined appropriations of state and federal governments for work within Colorado, excepting of course the U. S. Forest Service road activities. Every dollar which the State Highway commission has had at its command has been assigned to this working plan in order that full advantage may be taken of appropriations available to Colorado from the federal government upon a co-operative basis.

A glance at the sums assigned to the various projects in the five highway districts under this working plan gives an inkling of the great advancement in highway betterment which Colorado may expect in the next few years. The larger sums at the disposal of authorities have permitted larger apportionments to important projects than ever before possible. This means that many of the improvements which citizens in numerous communities have long desired will soon be on the way of realization. Work can now be done of which road enthusiasts have only dreamed hitherto. In light of the fact that similar appropriations, if not larger ones, will be available next year and the succeeding years the 1919 program is significant of a new era in Colorado highway development.

Naturally, it will not be possible to begin much of the work contemplated in this year's plan before fall or early next spring. However, some of the work will be actually in progress within several weeks as soon as contracts can be awarded.

In connection with this program it should also be remembered that the highway commission in January made appropriations totaling $500,000, which raises the grand total assigned this year to road work close to $3,000,000, exclusive of expenditures by counties on labors entirely local in character.

The largest amount of the apportionment just made is assigned to District No. 1, in which Denver is located, and logically, because the district comprises a greater population and greater present road mileage than any other one. The 1919 working plan, subdivided according to districts, follows:

DISTRICT NO. 1

(Total, $922,000.00)

Denver-Morrison, 3 miles	$ 75,000.00
Denver to hospital east, 2 miles	44,000.00
Denver-Brighton, 10 miles	230,000.00
Fort Collins, 3 miles	75,000.00
Greeley 3 miles	75,000.00
Platteville south, 2 miles	50,000.00
Longmont south, 2 miles	45,000.00
Loveland-Estes Park-Big Thompson	30,000.00
Boulder east, 2 miles	45,000.00
Morgan-Brush, 2 miles	45,000.00
Akron-Brush	48,000.00
Wray-Schramm	30,000.00
Morrison-Baileys	60,000.00
Sterling to Merino	70,000.00
	$922,000.00

Finish the plowing by packing with disc set straight

DISTRICT NO. 2

(Total, $357,000.00)

Cherry Creek road to Colorado Springs Junction (about 48 miles)	$ 35,000.00
Burlington north, 10 miles	15,000.00
Burlington west	10,000.00
Burlington east	5,000.00
Limon east, 10 miles	15,000.00
Colorado Springs-Canon City road	30,000.00
Colorado Springs-Palmer Lake	30,000.00
Colorado Springs-Cripple Creek	10,000.00
Castle Rock-River Bend	50,000.00
Cheyenne Wells, north or west	25,000.00
Woodland Park to Cripple Creek	25,000.00
Buena Vista-Salida	35,000.00
Buena Vista to Divide	40,000.00
Peyton-Ramah	17,000.00
Farmers' Highway, 18-s to Boyero	15,000.00
	$357,000.00

DISTRICT NO. 3
(Total, $404,600.00)

Trinidad-Walsenburg	$240,000.00
Prowers and Baca	50,000.00
Baca	10,000.00
Las Animas City east	35,000.00
La Junta west	11,000.00
Rocky Ford east	11,000.00
Manzanola west	23,600.00
Fowler east	24,000.00
	$404,600.00

DISTRICT NO. 4
(Total, $352,000.00)

Fort Garland to San Luis	$ 30,000.00
Monte Vista-Saguache	85,000.00
Silverton-Ouray	30,000.00
Top Norwood Hill to No. 7	50,000.00
Rico north	20,000.00
Delta-Montrose to Ouray County line	40,000.00
Delta-Hotchkiss	30,000.00
Durango-Mancos	40,000.00
Bayfield-Dyke	27,000.00
	$352,000.00

DISTRICT NO. 5
(Total, $306,900.00)

Steamboat Springs north, 15 miles	$ 59,400.00
Walden-Rand	30,000.00
Craig west to Maybell, 30 miles	75,000.00
Kremmling north	30,000.00
Meeker-Rifle	22,500.00
Grand Junction-Palisade	50,00000
Battle Mountain	40,000.00
	$306,900.00

If any of these projects do not meet the requirements for federal aid, the money will be returned to the contingent fund and reappropriated for other roads. These other roads upon which work may be done are: Whitewater to Gateway, DeBeque to Colbran, Grand Junction-Fruita, Craig west, Walden-Willow Creek road, State Bridge-Wolcott, Grand Junction-Delta, Glenwood to Aspen, Glenwood to Wolcott and the road north from Craig.—The Commercial (Denver).

Beet Top Silage

E. F. Rinehart

Field Animal Husbandman, University of Idaho

FOR a number of years the feeding value of beet tops has been recognized, hence some use has been made of this feed in the field. The method used, however, was wasteful for the reason that the tops were tramped in the dirt or dried up and blew away. As the livestock was allowed free access to the tops, undue quantities were eaten, resulting in digestive troubles as indicated by scouring, a decrease in the milk flow, or perhaps a loss of flesh. In case the weather was such as to cause decay of the tops there was danger of poison.

Since the value of succulence in the winter rations was recognized, considerable use has been made of corn silage and roots. In the beet growing districts of Colorado, Utah and sections of Idaho, some use has been made of beet top silage. While this is comparatively a new

---

A poor job of plowing is worse than none

---

custom in this country, the practice of making and feeding beet top silage has long been established in certain European countries.

Practically all the stockmen who have tried siloing the tops are enthusiastic and are of the opinion that one of the most valuable of the feed crops has been wasted.

It is from the experiences of these men that the following report is prepared. Mr. J. W. Jones and Mr. F. D. Farrell of the U. S. Department of Agriculture, and Mr. Mark Austin of the Utah-Idaho Sugar Company, who have made a study of the use of the beet tops as a winter feed, also furnished valuable suggestions.

II. Principles

The principles of storing corn silage successfully apply equally to beet top silage. The tops must be packed so as to exclude all air pockets and the outside air should be excluded as far as possible. Although it is considered desirable to run the tops through a silage cutter, they may be siloed whole if the mass is well packed.

III. The Silo

Undoubtedly the silos ordinarily used for corn would serve also for beet tops, but a less expensive type is used. The silo usually used is on the order of a potato pit.

Width—It is generally preferred to have the pit fairly narrow and to increase the capacity by increasing the length. In this way the surface exposed during the feeding period is minimized as the feeding is done from one end. The larger silos in use have mostly been from 10 to 16 feet wide at the bottom and from 12 to 20 feet wide at the top. If only a small amount of stock is to be fed they are made narrower. A Colorado report states that many of the smaller feeders are planning to use pits 4 feet wide at the bottom and from 6 to 8 feet wide at the top.

Length—The length is determined altogether by the amount of feed to be stored.

Depth—The depth varies from 3 to 6 feet. If the drainage conditions are right, 4½ or 5 feet is generally preferred. The tops are generally piled up 3 or 4 feet above the level of the ground, making the depth of the siloed tops from 7 to 10 feet.

The pit is usually prepared during spare moments. It is made with the ordinary slip scraper and the sides smoothed down and properly sloped by the use of spades or shovels.

If the character of the soil is not such as will drain readily the bottom of the pit should be made with a slight slope so as to facilitate drainage.

Both ends of the pit should slope so as to make it possible to drive in and out with the wagons while filling.

IV. Filling the Silo

When the beets are dug the tops are piled. As soon as possible they are hauled to the pit. If this is done within a few days they are but slightly wilted. In some cases the tops were not hauled until three weeks after piling, and though the silage seemed to be as good as when made from the tops that were put in within a week after harvest, the owners would not advocate this method unless labor conditions made it necessary. While there is some difference of opinion most men prefer to haul this within a week after the beets are dug. The majority of opinions were that from 3 to 5 days would be the most desirable time, although if the weather is good they might remain in piles for several days. In one case they were allowed to remain in the fields 40 days and although the feed was reported good the owner estimated that about 30 per cent of the feeding value had been lost.

In filling, the bottom of the pit is covered with about 18 inches of loose straw. On top of this the first layer of tops is placed. The wagons containing the tops are driven into the pits and unloaded so that in passing over the mass they assist in packing. It is recommended that the wagons are not driven in the same tracks each time as this would result in unequal compression which might be undesirable. Additional packing is accomplished in some cases by the use of horses and in others with heavy rollers.

The usual custom is to pack the tops as they come from the field, although a few make use of the silage cutter. If properly packed the feed keeps well either way.

When the first layer of tops has been well packed and has reached the thickness of 6 or 8 inches, salt is sometimes sprinkled over the tops. The amount used varies from 5 to 14 pounds to the ton of tops. Tops are added until there is a layer of 12 or 14 inches. Then a fresh layer of straw 5 or 6 inches thick is laid on and another layer of beet tops packed on. This process is repeated until the pit is filled and the top of the mass rounded off above the surface of the ground.

Some feeders do not use straw, however, the concensus of opinion is that where it is available it is well to use it as it reduces the washiness of the silage and facilitates the effective utilization of the straw which is eaten readily when handled in this way. The advisability of using straw seems to depend upon the freshness of the tops. Where the tops are gathered promptly before the moisture has evaporated from the leaves the straw takes up the excess moisture and the com-

---

Plowing can be done more cheaply now than in the spring

---

bination is desirable. If the tops are allowed to lie in the field until somewhat dry and then put into the silo with alternate layers of straw, it is more difficult to pack the feed properly, that is, the air is not excluded and the result is moldy feed. The proportion of straw used is about 1 ton of straw to 20 tons of tops.

When the last layer of tops is laid on, the final layer of straw is scattered over the top and the entire pile is covered with 1 or 2 feet of soil. The soil is placed on the mass to exclude air, shed water and prevent freezing, hence the thickness of the soil should be determined by these three requirements.

In filling the silo it should be remembered that the exclusion of air is a primary consideration. After filling is once begun it should be completed as quickly as possible to prevent spoiling. If the layers can be added gradually without any long periods when any particular layer is exposed to the air, the filling can be extended throughout the season.

V. Feeding the Silage

The pit may be opened at any time after filling, although many feeders prefer to have it closed for 30 days in order that the proper fermentation may take place.

One end of the pit is opened and the feed cut out in sections similar to the way hay is cut and fed from the end of a stack. Waste and spoilage is avoided by cutting only enough to feed in two days.

The quantities fed are approximately the same as for corn silage. Where fed with alfalfa hay a thousand-pound steer or a dairy cow will eat as much as 40 or 50 pounds a day, although less is usually fed, from 30 to 35 pounds being the average amount. Sheep are fed from 3 to 4 pounds each per day.

The tops should never be fed alone, but in connection with hay. When alfalfa hay and beet top silage are fed together a well balanced ration is provided. Where quick gains are desired a grain ration is fed in connection.

VI. Feeding Value

Practically all the men who have used this feed are enthusiastic over their results and are planning on putting up more this year. Most of the men visited have already prepared their pits for this year. Estimates of the feeding value vary from $5 to $10 per ton, while a few estimated its value equal to hay, others at half. The value of succulence to the winter ration is undoubtedly responsible for a part of the good results. Samples of the beet top silage were analyzed by the chemist of the American Beet Sugar Company and found to contain very nearly the same constituents as corn silage.

Beet tops as they come from the field contain beet leaves and the crowns of the beet roots. The crowns are rather hard and are sometimes difficult for the animals to swallow. As they come from the silo they have been rendered soft by the fermentation, hence are readily eaten.

If the tops have been put in fresh the silage is very succulent, which makes the layers of straw desirable and also makes it desirable to feed in connection with dry roughage. If fed alone it would prove too laxative for the good of the animals.

VII. Yield and Cost

The yield of tops per acre of beets is highly variable, the amounts varying from 4½ to 10 tons to the acre. The average of the reports is from 5½ to 6 tons to the acre.

The estimated cost of digging the pit, gathering the tops and filling the pit are practically all at $1 per ton. If the tops are purchased by the usual method of 25 cents per ton of beets this would bring the silage cost of $1.50 per ton. As before stated no one has estimated the value of the feed at less than $5 per ton.

VIII. Advantages

The following are a few of the advantages given by the men who have used the beet tops in this way:

1. The full value of the tops is secured.

2. Difficulties with choke and scouring are avoided.

3. A succulent, warm feed is provided for winter feeding.

4. Inexpensive and palatable feed is provided from two waste products, beet tops and straw.

5. Addition of the beet top silage to alfalfa hay is beneficial as forming a better balanced ration and also providing succulence, hence more profit is made from livestock.

6. The hay bill is reduced by one-half.   —The Utah Farmer.

Meteorological Report, Longmont

FOR MAY, 1918 AND 1919

TEMPERATURES:	1919	1918
Mean Maximum	86.50°	87.26°
Mean Minimum	46.00°	52.47°
Monthly Mean	66.25°	69.87°
Departure from Normal	+0.85°	−4.47°
Maximum	103.00° on 28th	97.00° on 14th
Minimum	30.00° on 1st	38.00° on 4th 29th
PRECIPITATION IN INCHES:		
To Date	3.51	8.22
For Month	1.18	0.86
Greatest in 24 hours	0.08	0.50 on 21st
Departure from normal	−1.28	−0.68
NUMBER OF DAYS:		
Clear	20	7
Partly Cloudy	9	22
Cloudy	1	1

Kansas Lamb Feeding Investigations

A. M. Paterson
Kansas Experiment Station

THE Animal Husbandry Department of the Kansas Experiment Station found lamb feeding to be a profitable venture during the winter months of 1919 in spite of the extreme high prices of feeds. We bought two hundred and forty-five Idaho lambs on the Kansas City market at $16.00 per cwt. The lambs were divided into seven lots of 35 each. The purposes of the investigation were to compare the efficiency and economy of linseed meal, cottonseed meal, and corn gluten feed as a protein supplement when fed with shelled corn, alfalfa hay, and corn silage; to determine the value of a protein supplement in a ration of shelled corn, alfalfa hay, and corn silage; to compare the efficiency and economy of shelled corn and hominy feed when fed with alfalfa hay and corn silage; to determine the value of adding corn silage to a ration of shelled corn and alfalfa hay, and to determine whether or not linseed oil meal can be used entirely to replace corn when alfalfa and corn silage are fed as roughage.

389

Lot I

	Feed Per 100 Lbs. Gain	Cost 100 Lbs. Gain	Daily Gain Per Lamb
Shelled Corn	$308.20	....	....
Linseed Meal	39.77	$15.02	0.4
Alfalfa	258.55	....	....
Corn Silage	377.87	....	....

Lot II

	Feed Per 100 Lbs. Gain	Cost 100 Lbs. Gain	Daily Gain Per Lamb
Shelled Corn	$386.02	....	....
Gluten Feed	49.81	....	....
Alfalfa	323.77	$18.74	.32
Corn Silage	473.19	....	....

Lot III

	Feed Per 100 Lbs. Gain	Cost 100 Lbs. Gain	Daily Gain Per Lamb
Shelled Corn	$360.38	....	....
Cottonseed Meal	46.50	$17.56	.34
Alfalfa	302.25	....	....
Corn Silage	441.76	....	....

Lot IV

	Feed Per 100 Lbs. Gain	Cost 100 Lbs. Gain	Daily Gain Per Lamb
Shelled Corn	$437.44	....	....
Alfalfa	366.88	$19.44	.28
Corn Silage	536.21	....	....

Lot V

	Feed Per 100 Lbs. Gain	Cost 100 Lbs. Gain	Daily Gain Per Lamb
Hominy Feed	$422.53	....	....
Alfalfa	368.01	$20.38	.29
Corn Silage	538.39	....	....

Lot VI

	Feed Per 100 Lbs. Gain	Cost 100 Lbs. Gain	Daily Gain Per Lamb
Linseed Meal	$362.73	....	....
Alfalfa	369.09	$19.61	.31
Corn Silage	566.36	....	....

Lot VII

	Feed Per 100 Lbs. Gain	Cost 100 Lbs. Gain	Daily Gain Per Lamb
Shelled Corn	$321.82	....	....
Alfalfa	555.40	$17.00	.38

	Average Weight Beginning Period	Average Cost Per Lamb	Feed Cost Per Lamb	Labor Cost Per Lamb	Interest on Investment and Equipment Per Lamb	Shipping and Market Expense	Total Cost of Feeding	Average Selling Weight	Average Sale Price Per Lamb	Net Profit Per Lamb	Selling Price Per Cwt.
Lot 1	64.56	$10.59	$2.96	$0.29	$0.23	$0.43	$14.47	$84.27	$16.01	$1.54	$19.00
Lot II	65.4	10.71	2.95	.28	.23	.43	14.60	81.14	14.81	.20	18.25
Lot III	66.08	10.82	2.96	.28	.23	.43	14.58	82.94	15.55	.82	18.75
Lot IV	65.73	10.68	2.70	.28	.23	.43	14.32	79.12	14.66	.31	18.50
Lot V	65.62	10.73	2.70	.28	.23	.43	14.60	80.00	14.72	.11	18.40
Lot VI	65.22	10.67	3.02	.28	.23	.43	14.63	80.61	14.31	.32	17.75
Lot VII	65.23	10.68	3.21	.28	.23	.43	14.83	84.11	15.85	1.02	18.85

Shelled corn cost $1.50 per bushel, linseed oil meal $65.00 a ton, cottonseed meal $65.00 per ton, corn gluten feed $63.00 a ton, hominy feed $60.00 a ton, alfalfa hay $30.00 a ton, and corn silage $8.00 a ton. The corn silage was made in August from corn injured by hot winds.

The results indicate linseed meal is more efficient than cottonseed meal, and cottonseed meal more efficient than corn gluten feed, as a protein supplement for fattening lambs when fed with shelled corn, alfalfa hay, and corn silage.

The addition of linseed meal or cottonseed meal as a protein supplement to a ration of shelled corn, alfalfa hay, and corn silage increased gains, reduced costs of 100 pounds of grain, increased selling price per hundred weight and the ultimate profits.

The addition of corn gluten feed as a protein supplement to a ration of shelled corn, alfalfa hay, and corn silage increased gains and reduced slightly the cost of gain, but failed to produce the desired finish and made less profit than when no protein supplement was used.

The substitution of hominy feed for shelled corn fed with alfalfa hay and silage increased slightly the gains, but increased the cost of gains; it also failed to produce as high a finish as did corn, making the ultimate profits less.

Corn silage added to the ration of shelled corn and alfalfa hay did not prove profitable, except when linseed meal as a protein supplement was fed with the silage, alfalfa hay, and corn; this combination gave the greatest profits.

Linseed meal substituted for corn and fed with alfalfa hay and corn silage produced slightly larger gains than when corn was fed with alfalfa hay and corn silage, but the linseed meal, alfalfa hay, and corn silage ration failed to produce the necessary finish and sold for so much less per cwt. that each lamb made a loss.

—From "The Breeder's Gazette."

VALUE OF DISC AFTER HARVESTING
From "Farm Engineering"

Farmers in every section of the country are coming to realize the importance of discing the harvest fields as soon after harvest as possible. Many of them even hitch the disc to the binder. The advantage of using the disc is that it conserves what moisture there is in the soil. Discing done at this time serves two purposes: It will loosen the surface soil to catch and retain the water, and second, it will kill most of the weeds. Very frequently the soil is so dry and hard at harvest that it becomes necessary to weight the disc in order to have it do effective work.

How to Meet the Water Shortage

(From the Utah Farmer)

DEVELOPMENTS during the past few days have emphasized the truth of previous announcements concerning the shortage of irrigation water, that must be reckoned with throughout the season.

The precipitation for the season has been below normal. The warm, dry winds early in the season evaporated the natural moisture from the soil at a rapid rate. Because of continued dry, hot weather, it has been difficult to make the maximum use of irrigation water. We have had dry seasons before this and we have also had winds that were bad in previous years, but we have had only few if any previous seasons when there was such a combination of negative forces to be dealt with as this season. The situation is critical! Corrective measures must

---

Double plow your alfalfa ground this fall

---

be applied! Community team work must be harmoniously and intelligently directed. Every water user must be made to realize the gravity of the hour. The welfare of thousands of homes and the prosperity of the commonwealth are at stake. The trump card will be played upon the farm and the water user will deliver the final decree. It is a personal problem requiring community team work and intelligent, broad minded attitude if Utah and Idaho are to escape an extremely unfortunate result when the autumn harvest period arrives. If your neighbor's barn was on fire every one in the community would rush to the rescue. Every one would help to the limit of their ability; yes, even more. A collection would be taken up to aid the neighbor in getting a start to recoup his loss. In this crisis, it is not the neighbor's barn that is on fire, it is even worse, it is his crop that is on fire. Are you going to respond in his hour of need? Perhaps, you, too, have a field with precious crops that are also on fire. Do you want the sympathy and active help from your neighbor? Of course you do! Then lets organize an emergency water brigade and go after the situation determined to win out against such serious odds.

Here is a brief analysis and a corrective program:

Under some irrigation streams they have no reserve reservoirs. There will not be a normal July and August flow of irrigation. Several streams that are ordinarily dependable are now flowing less water than is usually available in August. This means a shorter water supply at this time than common, and a more acute shortage will be felt within a short time. Remedy: Let every water user make certain that every minute of his water turn shall have his closest, personal attention; not a gallon shall run to waste. Let the water be used on the most valuable crops. Beets must have attention else the crop will be very short. Beets are one of the most profitable of our farm crops. Therefore, every beet grower must ignore traditional history which prompted him to not start irrigation until July. This will be too late if many fields are made to yield a profitable crop. Apply the irrigation water now. Then cultivate and then cultivate again in such manner that a well conditioned surface will be left to conserve the moisture. After a short time, the leaves will spread and cover the soil surface and thus aid in holding the moisture reserve. The soil is an excellent reservoir if sensible methods are used to conserve the moisture.

Then, too, there are irrigation reservoirs back up the canyons that ordinarily hold sufficient water to meet the late summer needs. In many instances, this reserve is being drawn upon far in advance of the usual time.

Again, we say, apply the irrigation waters in your regular turn and store this moisture in the soil with the crops and then cultivate and make the maximum use of the water as it is applied.

Communities should hold meetings and, by popular sentiment or otherwise, determine that water-logged pasture lands shall not squander the much-needed water when intensely cultivated, important cash crops upon which community welfare depends, shall perish. Priorities on water, of course, must be respected, but there must be neighborly adjustment.

If your sugar beets have not been irrigated, they are probably about as dry today as they ususally are the last of July. Therefore, apply July methods during June. DO IT NOW! Then conserve by proper cultivation and make war upon the weeds. Weeds are moisture parasites.

In our front page story of March 22nd issue was related how a community of farmers united efforts and saved 400 acres of wheat and 15 acres of barley. The community united their streams, irrigated by relays both day and night and made the highest and most profitable possible use of irrigation waters. The co-operative spirit directed their efforts and brought handsome results.

Disc between your grain shocks

Shock Threshing vs. Stacking

I N one county seat town in the wheat belt twenty new threshing outfits had been sold by the first of May, and other sales were in prospect. This shows the plans going forward to handle the wheat crop quickly. A number of these outfits are of small size, adopted to the large farm or a ring of small ones. The purpose is, as one finds by talking with the buyers, to get the wheat to market ahead of the great rush that seems sure to take place as soon as grain can be threshed. As there is a comparatively small storage capacity measured by the total yield, the resources of terminal elevators and farm bins will be taxed as never before, to say nothing of railroad congestion, which will accompany a glut of hurriedly marketed grain.

The plan to get the wheat off the farmers' hands as soon as possible may lead to considerable loss if the month following harvest proves to be a wet one. Shock threshing has the disadvantage that it often proceeds slowly, and the natural impatience of the farmers waiting their turn, leads to threshing when the grain is unfit to thresh. We have all helped to put grain through the machine when

---

Save moisture with a surface mulch

---

better judgment was decidedly against it. The man who has planned to thresh from the shock seldom stacks. He has helped the neighbors thresh from the field and waits the slow movement of the outfit in his direction, hoping that weather will favor him. With impatient farmers waiting and the machine men anxious to please, grain is sent to the bins so wet it must be constantly handled over to keep it from heating. As there is shrink in any grain from the time it is threshed, if threshed from the field, till winter, there is no incentive to hold the crop unless there is a price advance in prospect. In the present harvest this prospect is absent unless the government will establish a progressive price that shall cover all shrink, pay for storage and insurance. If this is not done we may expect such a congestion of markets as was never before known, and let a wet season for threshing ensue and loss in the shock will mount high.

Last year, a favorable one for threshing, the mills and elevators handled lots of grain that was threshed too soon or too wet. The same haste that was apparent then will be duplicated again on a much larger scale, with the added glut which more threshing rigs will precipitate. In many cases there is little or no storage room on the farms. The farmer expects to haul grain direct from the machine to town. Here is where there is possibility of much confusion. If home elevators are full, as they will quickly be, and railroads cannot supply cars owing to terminal embargoes, what advantages will the farmer gain from more threshing facilities. We may confidently expect that some of these machines will be idle days at a time, for the farmer cannot risk putting wheat on the ground. We do not think we have overdrawn the picture, and we know that such an event is almost certain to occur this season.

What is needed at this time is a return to the old practice of stacking, now more than half forgotten. This is one method of storage that every farmer can avail himself of without much cost. It is well-known that stacking gives a brighter grain, and wheat that has taken the "sweat" is safe to store in any quantity. In many neighborhoods one does not see a single stack of grain; all is threshed

Conserve moisture by discing before plowing

from the shock. As stacking seems to be a lost art on many farms, we cannot expect to see the trim stacks of our boyhood days, but most any farmer can put up a stack that will keep. Of course the header handles the bulk of the crop where wheat is the mainstay, but the cornbelt is in the wheat business as never before and the grain must be put into the bundle, as the header is not adapted to our acreage. In half the usual time the average farmer takes to go the rounds of the neighborhood in a shock threshing crew, his grain could be placed in the stack where it will be safe from loss, and where it can wait the coming of the machine till late fall if desired. Under normal conditions this article would not be called for, though every year in some states there is great loss from wet weather catching grain in shock. The impending harvest is going to rush the farmer as never before. A very valuable crop of wheat is at hand and the hope is that it will not be lost to the world by moisture in delayed shock threshing.—D. H., Kansas.

Disc Stubble to Prevent Crust Formation

We are now in the midst of the annual harvest and it beats all how much grain the binders are digging up even in this season of short water supply, both under irrigation and on dry land. If small grain stubble is allowed to stand after harvest without treatment the surface of the soil soon dries out and bakes so that plowing can scarcely be done, if at all. We all know how hard a Colorado stubble field becomes after the grain is cut and the rainy season is doling out limited moisture in deferred payments. Of course, rains occasionally come to soften this crust, but these cannot be absolutely depended upon, especially in a dry year like this when everything is out of whack. The formation of a crust can be prevented by thoroughly double disking after the binder. If well done, baking and clodding of the surface will rarely occur. Land so treated may be plowed and fitted for seeding when unfitted land cannot be plowed. Such land will also take up rain or irrigation water much better than undisked land. We remember one instance in which experiments were made on this feature of soil management. The land was disked after the binder and plowed about August 1st. A similar piece across a sixteen-foot roadway was not disked. The plowing was done at the same time. No rain fell from July 2 to August 31. From August 31 to September 5 three inches of rain fell. The land that had been properly tended was well wetted by this rain and fall crops started and did well. The undisked land was wetted only about six inches, not enough to make it possible to thoroughly break down the lumps.

Disking after the harvest is now being practiced by men who have the time to do it and it is certainly an important preliminary treatment of wheat land. The advantages of this work can be readily understood when we remember that usually at harvest the soil is free from weeds and in good condition to plow, although this year everything is out of whack because of the drouth, which has hardened the ground. Usually we have the rainy season to help out. The fellow who has a tractor can go ahead and do this disking. If a wet spell comes on the weeds grow and interfere with the plowing, especially the Russian thistle. What is worse yet they take away moisture which should be saved for the succeeding crop. The disking serves another useful purpose also in so thoroughly mixing stubble, straw and manure with the soil as to preclude any possibility of a break in the capillary tubes, which sometimes occurs when heavy stubble or manure is plowed under. —Denver Field & Farm, July 26, 1919.

> Finish the plowing by packing with disc set straight

A Labor Problem

From "The College Farmer"

FARMERS complain for about three to six months in the year of the great difficulty of securing labor. There are six months in the year in which, as farming is conducted now on most farms, they can find no employment for men even at board wages. No man can live in these days by working six months in the year, and the farmer should not expect to get efficient help for six months when he cannot give the man work for the other half of the year. There is a great demand for laborers during the haying and harvest seasons, but after the threshing is over the farmer is quite as anxious to get rid of extra labor as he was to get it at the opening of the season.

Manifestly, there is no way of solving this problem except by reconstructing our farming operations so that the farmer who needs labor can employ it the year around. This involves going in the livestock business; or, to put it another way, the farmer should regard himself as both a producer and a manufacturer. His business during the summer should be producing raw material, and in the winter manufacturing it into such condensed products as beef and pork, and butter or milk or cream all the year round.

Many farmers think they would be in an agricultural paradise if they had cheap and efficient labor all the year. Looking at man in his social and moral aspect, we think he would be in anything but an agricultural paradise. For if labor were cheap and abundant, the small farmer would absolutely be put out of business. The capitalist would enlarge his holdings, would organize his farming business just as business is organized in large manufacturing concerns, and the small farmer would be forced to quit. More than that, there would be a deterioration in society that would make the country anything else than a good place to rear a family. There would be slums all over the country, as there are in the cities.

We hear a great deal about intensive farming, by which is meant putting a large amount of labor on a few acres. That means the development of a peasantry, whose life is one of the narrowest, and the social conditions of which are anything but desirable. We believe in intensive farming in this sense: that the farmer should put sufficient labor on his land to grow crops at the lowest cost. We do not believe in an expenditure of labor on land to produce the highest possible yield without regard to cost.

As Professor Carver recently said in an address at Manhattan, Kansas: "Wherever employes are scarce and hard to find, social conditions are good. Where employes are abundant, social conditions are bad." This is absolutely true in the cities; and if labor were as easily available in the country as in the city, it would be quite as true there. Therefore, the high price of labor is not an unmixed evil, but has a great deal of good in it.

The solution of the problem is plain and easy, and is being solved by advanced farmers today. It is simply this: Arrange to employ your labor the year around and reduce the amount of extra labor during harvest to the minimum. This is entirely practicable, but involves the reconstruction of our system of farming, a reconstruction which we will be forced by circumstances to do.

Official Investigation Shows Value of Tractor

(From the Utah Farmer)

University of Idaho recently completed an investigation of the use of tractors on farms in that state.

Since this was an official and unbiased investigation of the results being obtained with tractors in actual use on farms, it is believed the summary as published by the University may be of interest to farmers. We are therefore quoting it below:

"1. The tractor, when selected to suit the farm and intelligently and carefully operated, has been reported by owners in Idaho a profitable investment.

"2. The farmer must be able to make all minor repairs himself and get repairs and expert help quickly for larger installations.

"3. Dependability is probably the largest factor in the success of the tractor.

"4. The three-plow size is favored by a majority of Idaho owners.

"5. Proper care of lubrication will prolong the life of the machine.

"6. The best quality of oil is the cheapest for the tractor.

"7. The tractor motor is required to pull its rated load the greater portion of the time. The automobile motor is rarely subjected to full load for a ten-hour day. Care for the tractor accordingly.

"8. The tractor displaces half its value in horses can easily be made to pay its way, according to reports of Idaho tractor owners.

"9. The man who makes up his mind to care for his machine and to be as independent as possible of outside help is the satisfied owner.

"10. Taking off one plow may enable the tractor to operate at its rated speed and may result in a greater accomplishment for the season.

"11. Over loading brings trouble and cuts down the work accomplished during the season."

The author of the bulletin in which the results of the investigation were published also offered a few "Hints to Prospective Owners" which we believe should have careful consideration by all prospective tractor farmers. They are as follows:

"Know your tractor thoroughly before you begin operating it. Study your instruction book and know the why and the how of each part of the machine.

"Follow instructions of the manufacturer. He has experimented with cheap oils and with heavy loads and in your instruction book you are getting the benefit of his experience. The experimental work is expensive. Let the manufacturer do it.

"Spend as much time morning, noon and night on your tractor as you would on your horses. Keep your whole ignition system free from dirt and grease. Clean all working parts that are exposed to dust. Grease and dust together make a very effective grinding compound.

"Do not overload the tractor as it will shorten its working days and bring you expense.

"House the tractor between seasons. Proper shelter will be far cheaper in the long run."

—From "The Utah Farmer," June 28, 1919.

THROUGH the LEAVES

SEPTEMBER, 1919

THE GREAT WESTERN SUGAR COMPANY

THROUGH
THE
LEAVES

SEPTEMBER, 1919

Published Monthly By

THE GREAT WESTERN SUGAR CO.

DAILY CALL PRESS
LONGMONT, COLORADO
1919

TABLE OF CONTENTS

Editor's Page

N. R. McCreery

IT is an evidence of a convinced community to see the very large number of farmers who have gotten their grain out of the way and started to turn their ground over for next year's crops. The early planting of beets this year has meant more perhaps than most years, as early moisture brought up early planted beets. Fall plowing means early planting. We believe any farmer is justified in having his ground plowed in the fall even if he has to hire it done.

* * *

The landlord is certainly interested in the yields from his land. This is true even on land that is rented for cash, as land that does not produce well means a low rental value. This should mean that the landlord is interested in making a lease that will encourage fall plowing and fertilization. The one-year tenant system is one of the worst things in any farming community, as it means robbing the soil and usually no systematic crop rotation. Take note of the successful farms around you and see if there is a new man on them each season. Make it worth their while to stay on the place. They can do much better for you the second year than they can the first, as they are better acquainted with your farm.

This matter of tenants is so important to both parties that we have deemed it worth while to secure a few articles from those who know. These will appear during the next two or three issues and we hope will help in correcting this serious condition that keeps yields down on many farms.

* * *

We have learned of a number of growers who are going to silo their beet tops this year for the first time. Certainly $25.00 per acre is better than $8.00 or $10.00 per acre. Besides the feeding of the siloed tops will make more fertilizer than the pasturing in the field. Make your plans now to silo your tops this year and realize this full value. In that way you will be able to feed your pulp in the earlier part of the season and can be sure of your full allotment.

* * *

Some growers are desirous of having their beet tops left somewhat in piles to help them in piling them. The Mexican beet help will probably want to leave them in rows. We believe the labor

405

contract is plain that the beets must be thrown in piles before topping, if the grower desires. Call your fieldman if you have any difficulty in getting them to do this.

* * *

Beet harvest will probably begin some time the last of September. Do not wait until you receive your harvesting orders to get your plows and wagons ready. Your neighbors will get the start of you if you do. Also there are always so many who wait till the last minute that the supply of blacksmiths and repair men is too small to do the large amount of work all at once. Be prepared ahead of time and keep ahead.

* * *

CORRECTION NOTICE

The meteorological report published in the last issue was for June, 1918-1919, instead of May.

In the table on page 390 of August issue, under the caption "Feed Per 100 Lbs. Gain," the figures used represent pounds, not dollars.

Meteorological Report, Longmont

FOR JULY, 1918 AND 1919

TEMPERATURES:	1919	1918
Mean Maximum	91.13°	83.32°
Mean Minimum	55.22°	54.26°
Monthly Mean	73.17°	68.79°
Departure from Normal	+3.27°	−0.91°
Maximum	100.00° on 23, 24, 26	96.00° on 2, 31
Minimum	48.00° on 15th	45.00° on 1st
PRECIPITATION IN INCHES:		
To Date	5.00	13.23
For Month	1.49	5.01
Greatest in 24 hours	0.67	1.63 on 9th
Departure from normal for mo.	−0.42	+3.07
Departure from normal since Jan. 1	−5.38	+2.53
NUMBER OF DAYS:		
Clear	17	6
Partly Cloudy	11	24
Cloudy	3	1

Finish the plowing by packing with disc set straight

Notes

H. Mendelson

Education by Violence.

The writer recently made a trip through the various sugar beet districts between Idaho Falls, Idaho, and Toledo, Ohio, including some parts of Wisconsin. Everywhere the effect of the drouth was painfully visible; and yet everywhere a few beet fields could be found ot very much better appearance than the suffering fields, although they had no more rain than the others. In irrigated districts, usually it is assumed a field above the average had some water applied to it. In the districts relying entirely on rain, this, of course, cannot be the case.

While many factors affect the ultimate yield of a beet field and a superficial inspection does not always reveal all that it is necessary to know, still the statement can be made that generally the earlier planted beets, east or west, are the best. There are not very many cases on record where a farmer makes a mistake by doing things too early. Most mistakes are made by doing things too late.

Naturally the effects of too late planting are more visible in an abnormal year like this. But they exist in a normal year just the same. In some parts of the East they seem to trust a great deal in the longer growing season, they claim. In our irrigated section there certainly is not a long growing season. The only way we can make the growing season longer is by giving the plants an early start by any means feasible and available. We hope that the lessons of the violent drouth will be remembered for a few years anyhow.

The Greek historian, Thucydides, said some 2,300 years ago, "War is education of violence." Moral suasion is supposed to be the modern method by government and education. Still, cities seldom start a campaign of sanitation except after an epidemic. In other words, locking the barn door after the horse has escaped is practiced in cities as well as on farms.

* * *

Learning to Fight Webworms.

One of our managers reports on his campaign against webworms as follows: "The outbreak last year was largely in the J————— and M————— districts, and farmers there were willing to spray promptly, as they were wide awake to the danger; but in the rest of the districts it was hard to get growers to realize the danger and some of them had to see the damage the worms could do before they woke up." Evidently education by violence is the preferred method.

Why the Ditches Were Not Ready.

In a very fine irrigated district outside of Colorado, people finally awoke to the necessity of irrigating much earlier in the spring than usual. The river contained an ample supply of water, but the ditches were not cleaned out yet. In this particular wealthy farming district water is kept running in the ditches all winter to furnish the supply for man and beast. No cisterns exist. So the first thing on a cold winter morning the lady of the house does, is to make a trip to the ditch, chop a hole in the ice and fill the time-honored bucket. Therefore, the ditches are cleaned late in the spring, when the spring rains fill the rain barrel in normal seasons. There will be many cisterns built in this district during the summer.

* * *

Wisdom of Sixteen Years Ago.

We read recently a pamphlet printed in 1903, entitled "A Practical Talk to Practical Farmers on Sugar Beet Culture," by C. S. Faurot, then Agricultural Superintendent of the Longmont Sugar Factory. Among other good things the writer says: "Beets should be planted as early in the spring as possible. Experience has shown that the early planted beets produce a larger yield and a higher sugar content than the late planted. * * * Experience has shown that the beet seed can be planted with safety in Colorado as early as the first week of April, when the soil and climatic conditions are favorable. * * * The season of planting should not extend beyond the 15th of May. * * * As it is very important that a good stand be secured, I would suggest that should the ground be dry when seed bed has been prepared, the ground be laid off in rows from 2 to 3 feet apart. This can be done very quickly by taking three or four round posts about 3 feet long by 6 inches in diameter, nailing a plank on top of the posts and furrowing out the field in the same direction as it is to be planted, then giving it a light irrigation. * * * Let me caution the beet grower against the false idea that I find prevailing, that is, if the young beet be allowed to suffer for the want of moisture, it will seek the moisture below and therefore penetrate the ground to a greater depth than it otherwise would, had the surface been irrigated. * * * Never let the young beet suffer for water to the extent that the foliage wilts and goes down into the ground, for when that condition prevails, your beet has not sufficient moisture to long sustain life, much less to keep it in a good healthy growing condition, for when a young beet has once been checked in its growth to any very precipitated degree, either from want of moisture or any other cause, it becomes a stunted, puny plant and will never mature you a good crop."

How much have we learned since 1903?

* * *

What They Do Better in the East.

The absence of irrigation ditches overgrown with weeds, crossing the fields at irregular angles, gives the farms in the corn belt a much neater appearance. Their method of cultivating corn is fully equal in effectiveness to our average beet cultivation. The farms are cropped to the fence, as they can plant their rows parallel with the fences. A little more attention to this could be paid on our irrigated farms. One does not realize how much land in our farms remains uncropped, due to the necessities of irrigation, until one sees these good eastern farms.

During the last half of July in quite a number of fields from eastern Nebraska to Toledo, Ohio, manure was being spread on the green stubble, in some instances where the shocks were still standing. Plowing had commenced. In the beet districts north and south of Toledo and in northwestern Wisconsin, virtually all land intended for beets is fall plowed.

* * *

What We Try to Do on Our Farms.

In this connection, the following part of a letter sent by Mr. Comer, August 8th, to our managers with reference to fall plowing on the Sugar Company's farms may be of interest to our farmers:

"Plans should soon be made for the fall plowing campaign. In laying out the work of harvesting and threshing this year, we have endeavored to arrange the several operations in a manner to make it possible to push the fall plowing through without unnecessary delays. Include at least, all lands required for seed beets and stecklings next year, and as much as possible of the land intended for planting to other crops next spring should also be fall plowed. It pays to follow this practice even if it is necessary to hire outside teams or tractors to do so.

"All grain and beet seed stubble land should be double disced before it is plowed, and the earlier this discing is done the better. If at all possible, it should be done before the shocks are taken out of the fields. Whenever possible double disc after plowing. Throughout our whole district there has been a water shortage this year, and as a consequence the subsoil is abnormally dry. If the water supply will permit your doing so, after the land has been fall plowed, irrigate it, using deep furrows spaced about two feet apart. We specify deep furrows to insure against flooding and to prevent blowing of soil during the winter.

"Regarding Plowing Alfalfa Land. The method to be followed in plowing alfalfa land will depend on local conditions to a greater extent than in the case of plowing stubble land. If it is desired that hay be made of the third cutting, and if time wil permit, crown it as shallow as possible immediately after the third cutting is taken off. Spring tooth harrow immediately behind the plow; and then follow up later by replowing seven to eight inches deep. A thorough job also calls for double discing after the replowing, and if you can fall irrigate, this should be done too. If time will not permit plowing twice a short-cut method can be used by plowing once seven to eight inches deep. If this method is adopted it is more important to double disc after the plow than if the field is first crowned and then replowed. This kind of land should be irrigated, if at all possible, before it is plowed to help kill the crowns as explained later.

If the alfalfa stand has not been in long enough to put the soil in the state of fertility desired, provided the third cutting is not used for hay, and if you have a sufficient water supply to irrigate before plowing, then we recommend plowing under the third cutting (especially if the third cutting will not produce a normally heavy yield). Plow about eight inches deep, using a chain or some other device to turn the green crop entirely under. Double disc after the plowed land has dried out on top. Irrigate before the land is plowed, and plow as wet as possible, even though it turns up slick. Sufficient moisture must be provided to assure that all the alfalfa will decay instead of drying out. Decay kills the crowns. If a third cutting is turned under in dry soil and it remains dry, the benefit of the operation will be greatly discounted, as plants will not decay and next spring many of the crowns will grow. Trash of this strawy nature and sprouting crowns do harm to a seed bed, instead of accomplishing the benefit expected at time of plowing alfalfa crop under."

* * *

What They Don't Do As Well in the East.

In our climate it is thoroughly understood, and often practiced, that thorough cultivation is a very essential help in preserving moisture. In eastern regions where they suffer more frequently from too much moisture, deep cultivation on the beet fields was mostly noticeable by its absence. One exception was a field on a farm operated by the Sugar Company showing all the evidences of good cultivation, and the crop showed it. It was the best looking field in the whole territory, and suffering as yet very little from the drouth.

FULL VALE FROM ENSILAGE
Cutting Before Soft Dough Stage Is Reached Is Not Good Practice
(Prepared by the U. S. Department of Agriculture.)

Cutting corn for the silo before it is ready is too general a practice. The full feeding value of the crop cannot be obtained until the soft dough stage is reached. The lower leaves will then be dead, some of the husks will have turned brown, and the ears will be hard, but the stalks and upper leaves of the plants will still be green and succulent. Cutting before this time is like marketing cattle or hogs before they are finished.

—Denver Field and Farm, Dec. 30, 1918.

FINISH THE JOB

A. C. Maxson

The beneficial effects of fall plowing are often missing because the plow is not followed by the harrow, disk or roller as conditions may require.

The theory that fall plowing should be left rough to catch and hold moisture is all right when moisture comes, however, in case of a dry winter the clods fail to become moistened and are not broken up by frost. When the spring is dry a seed bed of clods to a depth of several inches deep usually results. Poor germination of seed, a poor stand and a poor crop often follow.

If the soil is reasonably well broken up by the plow, use a harrow, if it is cloddy, use a disk set straight, and if still cloddy, force the clods into the plowed soil by rolling. Do not be afraid of getting the soil too firm. Firm ground freezes more deeply than loose ground. The freezing will loosen the soil before spring.

Don't spoil a good job of fall plowing by not doing a little more work in the form of harrowing, disking or rolling before it freezes up.

Harvest Losses

A. C. Maxson

IT does not matter whether a farmer is engaged in **stock** raising, raising general farm crops or in some one **of** the more specialized types of farming there **are** certain losses which, in the course of time, he is called upon to bear.

Some of these losses are unavoidable while **others** can and should be avoided. We soon reach a point where we take the unavoidable losses as a matter of course **and** cease to worry over them. There is also danger of looking at the avoidable losses with the same indifference with which we view the unavoidable ones thereby suffering needlessly.

As the beet harvest approaches we are reminded that every year many growers suffer losses which can and should be avoided **or at** least reduced.

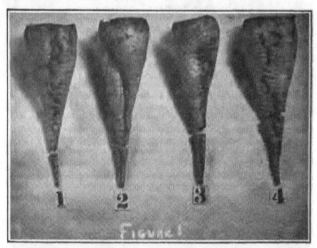

Large and visible losses are usually given prompt attention and quickly remedied. It is the small, invisible leak that is not noticed and which continues to levy its unsuspected toll.

A common though unappreciable loss is due to breaking the tap root of beets by using an improperly adjusted beet puller.

Frequently a beet grower attempts to run a puller shallow in order to make it possible to use one less horse. Again the puller is run on the points in an attempt to force it into hard soil. In both cases many beets are broken off nearer the crown than they should be. The loss thus incurred is not noticeable but nevertheless it is worthy of attention.

Fig. 1 represents four beets broken off at varying distances below the crown. In all cases the entire beet as shown should be removed from the field. Beet No. 1 has one-quarter ounce broken from the tip; No. 2, one-half ounce; No. 3, one ounce and No. 4, one and one-half ounces. The following tables show the loss in pounds beets and the money loss per acre represented by each beet on a 12, 14, 16, 18 and 20-inch stand:

Beet No. 1—One-Quarter Loss Per Beet

Avg. Stand	No. Beets Per Acre	Lbs. Beets Left in Ground	Loss Per Acre. at $10.00 Per Ton
12-inch	26000	406	$2.03
14-inch	22400	350	1.75
16-inch	19600	306.2	1.53
18-inch	17400	272.8	1.36
20-inch	15700	245	1.22

Beet No. 2—One-Half Ounce Loss Per Beet

Avg. Stand	No. Beets Per Acre	Lbs. Beets Left in Ground	Loss Per Acre at $10.00 Per Ton
12-inch	26000	812.5	$4.06
14-inch	22400	700.0	3.50
16-inch	19600	612.5	3.06
18-inch	17400	543.7	2.71
20-inch	15700	490.6	2.45

Beet No. 3—One Ounce Loss Per Beet

Avg. Stand	No. Beets Per Acre	Lbs. Beets Left in Ground	Loss Per Acre at $10.00 Per Ton
12-inch	26000	1625	$8.12
14-inch	22400	1400	7.00
16-inch	19600	1225	6.12
18-inch	17400	1087.5	5.42
20-inch	15700	981.2	4.90

Beet No. 4—One and One-Half Ounce Loss Per Beet

Avg. Stand	No. Beets Per Acre	Lbs. Beets Left in Ground	Loss Per Acre at $10.00 Per Ton
12-inch	26000	2437.5	$12.18
14-inch	22400	2100	10.50
16-inch	19600	1887.5	9.18
18-inch	17400	1631	8.13
20-inch	15700	1470	7.35

413

It will be noticed that if we have a 20-inch stand and lose one-quarter ounce per beet (see beet No. 1) the money loss per acre is $1.22. On a ten-acre contract the loss would be $12.22, a twenty-acre contract, $24.44, a thirty-acre contract, $36.66 and a fifty-acre contract, $61.10. These losses are occasioned by breaking a piece of beet two inches long and three-eighths of an inch in diameter at the largest end from every beet.

If we break a piece four and one-half inches long and one inch in diameter at the largest end (see beet No. 4) from each beet, the losses for the same sized contracts and a 12-inch stand would be: a ten acre contract, $121.80; a twenty-acre contract, $243.60; a thirty-acre contract, $365.40; and a fifty-acre contract, $609.00.

These represent the extreme losses represented by beets Nos. 1 and 4. As a matter of fact the actual loss in any field would approach the average of these two more nearly than either of the extremes. The average loss represented by the tables given above is $3.12 per acre and should not be far from the average loss of all fields under normal harvest conditions. For a ten-acre contract this would amount to $31.20; for a 20-acre contract, $62.40; for a 30-acre contract, $93.60; and for a 50-acre contract, $156.00.

Another loss is due to leaving whole beets in the field. This in turn is due to improperly lifted beets, careless hand labor and tramping by the horses on the puller.

Beets which are not properly lifted will be left by the hand labor.

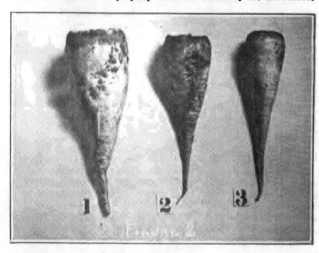

In case of a wet fall many beets are forced into the soil by the horses stepping on them. This can be remedied to some extent by having the eveners of proper length and the cross reins properly adjusted.

Fig. 2 represents three beets—No. 1, weighing two and one-half pounds; No. 2, one pound; and No. 3, three-quarters of a pound.

If your beet rows are 20 inches apart, there is a total of 26000 feet of beet row per acre. Sixteen and 24-inch rows and 18 and 22-inch rows alternating have 26000 feet of row per acre also.

Supposing we leave a beet like No. 1 in every 100 feet of row; we leave 260 beets per acre. These will weigh 650 pounds. At $10.00 per ton they are worth $3.25.

Figuring in this way we find the loss in beets and money where beets like each of those in Fig. 2 are left every 50, 100, 200, 300, 400 and 500 feet in the row. Harvest Losses

Beet No. 1—Weight, 2½ Lbs.

	Beets Left Per Acre	Lbs. Per Acre	Loss Per Acre at $10.00 Per Ton
1 beet every 50 ft.........520		1300	$6.50
1 beet every 100 ft.........260		650	3.25
1 beet every 200 ft.........130		327	1.62
1 beet every 300 ft......... 86		215	1.07
1 beet every 400 ft......... 65		162.5	.81
1 beet every 500 ft......... 52		130	.65

Beet No. 2—Weight, 1 Lb.

	Beets Left Per Acre	Lbs. Per Acre	Loss Per Acre at $10.00 Per Ton
1 beet every 50 ft.........520		520	$2.60
1 beet every 100 ft.........260		260	1.30
1 beet every 200 ft.........130		130	.65
1 beet every 300 ft......... 86		86	.43
1 beet every 400 ft......... 65		65	.325
1 beet every 500 ft......... 52		52	.26

Beet No. 3—¾ Lb.

	Beets Left Per Acre	Lbs. Per Acre	Loss Per Acre at $10.00 Per Ton
1 beet every 50 ft.........520		400	$2.00
1 beet every 100 ft.........260		195	.975
1 beet every 200 ft.........130		97.5	.487
1 beet every 300 ft......... 86		64.5	.32
1 beet every 400 ft......... 65		48.75	.24
1 beet every 500 ft......... 52		39	.195

The smallest loss, 19½ cents, represents a three-quarter pound beet every 500 feet. This appears too small to bother with. Yet how many of us would not take the trouble to pick up a penny. A little more supervision of the hand labor, a little more attention to the way our puller is working and the way the horses are hitched will save this 19½ cents and cost us nothing.

The average loss represented by the foregoing tables is $1.30. This is the equivalent of leaving a one-pound beet every 100 feet. For a 10-acre contract the loss would be $13.00; a 20-acre contract, $26.00; a 20-acre contract, $39.00; and for a 50-acre contract, $65.00.

Is it not worth while to spend a little more time supervising the harvest and adjusting the puller in order to prevent such losses?

Values

The question of the value of beet top ensilage has been raised this year because of the high price of other feeds. Hardly a day passes that there isn't some inquiry made as to beet top ensilage.

The following table has been compiled to show the comparative values of some feeds with ensilage based on the present market prices:

	Assumed Price Per 100 Lbs.	Lbs. Beet Top Ensilage Required to Replace 1 Lb. Feed	Value of 1 Ton Silage Compared With Feeds
Wheat	$ 3.30	6.5	$10.16
Corn	3.90	7.7	10.14
Barley	3.15	6.5	9.68
	Per Ton		
Alfalfa	$15.00	2.3	$ 6.51
Molasses	20.00	4.5	4.44
Corn Silage	5.00	1.17	4.27
Pulp	1.25	.43	2.90

It is very clearly shown that beet top ensilage is a feed well worth considering if you are contemplating feeding this year.

Farmers who silo their tops find it is better not to let tops lay on the ground more than two or three days before siloing, and for that reason the silo should be built before the beet harvest starts.

Double plow your alfalfa ground this fall

Fall Plowing

J. F. Jarrell

THE benefits to be derived from fall plowing are many and too obvious to some farmers to need any mention.

As we plan our fall work we should include every possible day to get our ground plowed for the next year's crop. When spring comes and our fields are ready to be seeded we are a great many jumps ahead of the fellow who has neglected to fall plow and has all of the work to do to get his seed beds ready.

This year should serve as an object lesson to us. The dry spring which handicapped us in so many ways and caused so much loss of crops emphasizes the importance of early planting. Early planting requires early seed bed preparation and nothing is more conducive to early final work on a seed bed than beginning in the fall by plowing.

Some soils blow badly if plowed in the fall. This can be prevented to a great extent by leaving the surface rough or by corrugating as is practiced in some sections.

A thorough disking of the surface of the soil is a valuable practice previous to plowing. This makes a well tilled layer which when turned under goes a long ways towards fitting the lower layers of the seed bed closely together, and avoids the making of air pockets which cause so many short and sprangled beets.

When a farmer wishes to plow more deeply than his field has been plowed, the best time to do it is in the fall. The new soil which he turns up will then have a chance to weather down and be in condition to be worked into a good seed bed early the next spring. Fall plowing makes early planting. Early planting makes more tons of beets. More beets make more dollars, and after all it is the dollars that we are growing, even though we have to handle some crop to get them.

A farmer might well afford to hire extra teams in order to get his ground plowed in the fall. If he has mares which are being used as breeders, he can get some of the expense of their keep by plowing with them instead of letting them loaf during the fall months.

Of course where a beet crop will follow beets, it might crowd the work a little, but by using the two-way plow and siloing the beet tops the plowing can follow up the digging closely so that a few days after digging the field will be ready to go into winter in first-class shape for the finishing touches to be made early next spring. This will permit the seeding of beets and spring grains and the planting of other crops earlier, which in most cases means better yields and finally it is the yields (the dollars) that we are working for.

Let us sharpen our plow shares and get busy. Let's go from here!

417

Farm Superintendent Believes in Fall Plowing

(We are printing below a letter received from Mr. John Maier, General Farm Superintendent of the Billings district, in answer to our request for an article on fall plowing. We think it merits being printed in "Through the Leaves.")

August 26, 1919.

Mr. N. R. McCreery, Manager,
The Great Western Sugar Co.,
Longmont, Colorado.

Dear Sir:

I have yours of August 22nd requesting an article for "Through the Leaves" on reasons fall plowing has not proven successful in the Billings District.

I am afraid you will be disappointed because I am a booster for fall plowing and believe it has been proven successful. It has been my hobby in season and out of season; if ever there was a district, in my mind, where fall plowing is needed, it is the Billings District and Big Horn Valley. Our seasons are short at best and any work accomplished in the fall means that much gained in the spring which adds days to the growing season. Our soils here are extremely heavy; a majority of it is heavy gumbo and let me assure you that if any soil needs fall plowing it's the gumbo soils.

Practice spring plowing on this kind of soil and you have, as a rule, one of the worst conditions imaginable. It is utterly impossible to prepare a decent seed bed. Fall plowing, however, followed by any reasonable amount of winter moisture leaves this kind of soil in the most desirable condition. Take a lump of gumbo and add a little water and expose to the air and you will get about the same results you would get if you poured water on a chunk of quick-lime, except that the action is slower, but the result on gumbo soil is a perfect seed bed and one of the greatest advantages is that soil in this condition will not crust, and talk about your mulch,—at Hardin where we had some fall and winter moisture, we had ideal conditions and a perfect mulch and good seed beds were the rule and practically all that is needed to be done in the spring is to level the ground, harrow and plant.

West of Billings and in the Clarks Fork Valley, conditions were altogether different, last winter was extremely dry, much of the fall plowing was too dry for good work. Still we all figured dry, lumpy fall plowing better than no plowing at all and consequently a large amount of acreage was fall plowed. An open winter with frequent winds took every bit of moisture out of this soil. Spring rains failed and this left all fall plowed fields in bad condition. No doubt much of this plowing could have been worked down with disc and harrow and some moisture might have been saved, but who ever heard of winter moisture being short in Montana, such a thing was thought to be impossible and of course working down soil in the fall means work which the elements can do better and ever so much cheaper, but believe me here is one who is not going to be caught napping again. It's too darn costly; we cannot afford to take the chance. A little extra work on the soil in the fall is like carrying insurance against fire.

Well, I have rambled a lot and said very little. I hope you will throw this in the waste basket, but I am convinced that fall plowing is essential for this country and working down the soil immediately after is the only safe practice.

Very truly yours,

JOHN MAIER.

WHY KEEP FARM ACCOUNTS?

The Mountain Farm Management Association, which is connected with the state college, is offering handsome cash prizes, donated by various business firms throughout the state to the farmers keeping the best and most accurate set of farm records and to those having the best organized farm as shown by the records. Prizes are also offered to the farm bureau in the county having the largest number in the contest.

The object of this contest, namely, to stimulate interest among the farmers in keeping a business record of their financial affairs, is meritorious. Farmers nowadays must learn to be business men. They cannot operate intelligently and expect to make progress unless they keep accurate, complete and up-to-date financial records. This really takes very little time—but no matter how much time it takes, it is worth it. Only in this way can the farmer know exactly what branches of his farm operations are making him money, what ones, if any, are losing him money, and how much. Then he can put his finger on the weak spots.

The Saturday Evening Post in a recent issue hits the nail on the head when it says that the farmer who does not keep farm accounts is as far behind the times as one would be now trying to plow with a wooden plowshare. And one of the best reasons why the farmer should keep a set of accounts is that it improves his credit and increases his borrowing capacity. To the banker it is an evidence of responsibility and progressiveness. —Midland Empire Farmer

Disc between your grain shocks

Sugar Beets in Southwestern Russia

Lieut. Nicholas Kobliansky

IN the July number of "Through the Leaves" there appeared a short story under this same heading about planting and cultivating sugar beets in the southwestern corner of European Russia. Now, on acount of the forthcoming harvest of beets in northern Colorado, it will be interesting for the readers of "Through the Leaves" to read about the digging of those beets in Russia. According to the contract a grower is obligated to begin digging by the first of September, and from this time on the beet fields are again covered with gangs of diggers working with spades. Every party of diggers, generally consisting of the members of one family, works a strip across a field the width of which depends upon the number of workers in the party. The beets which have been dug are thrown into small piles where, with a knife six to eight inches long, the green tops, roots and dirt are removed. A special inspector watches for clean digging without breaking or missing beets; regular cleaning; and accepts cleaned beets by measuring them with a special box without a bottom which holds about one ton of cleaned beets. This box is filled by baskets from the small piles left along the strips by the diggers. The box has four handles and after it is filled, it is lifted, leaving the beets. The beets must then be immediately covered with beet tops and dirt. Such piles follow in rows on every strip. All beets dug during a day must be cleaned, measured and covered in piles before night. To leave cleaned beets in the field uncovered over night is not allowed. Every such pile as described which is accepted by an inspector is paid for with about 40 to 50 cents.

With the beginning of harvest delivery of the beets to the sugar factory also begins. Deliveries are made in regular vehicles which will carry one of the described piles. These wagons belong to a grower or are hired peasants' wagons from the nearest villages. Hundreds of these wagons are moving on the roads leading to the sugar factory from all directions. Every party of such wagons arriving at a sugar factory is accepted by the sugar factory's inspectors and scale men by weighing two or three wagons from every party containing generally 30 to 50 wagons. The beets are accepted by this weight less 5 to 20 per cent for dirt, according to the estimate of the inspectors. With an average of about six miles from the beet field to the sugar factory and all the procedure of acceptance, it takes one day's work for a team to deliver one ton of beets to the sugar factory. Such a wagon, with a team and man, was paid at the rate of from $1.00 to $1.25 per day. Such was the work and such were wages in 1914-1915.

I don't think that any of the farmers of northern Colorado would grow sugar beets under such conditions. How far those conditions are from improved machines, cultivators, pullers, etc.; and what a difference there is between hauling one ton of beets a day 6 to 10 miles then shoveling them on high piles around the factory and dumping beets not farther than a mile from a beet field over the excellent dumps of the Great Western Sugar Co. And last, how big is the difference between 3½-4 dollars for a ton of sugar beets perfectly cleaned and piled on the field with all the cheapness of help and 10 dollars a ton for beets delivered to closely located dumps immediately after digging and cleaned only from green tops.

Colorado farmers may say, "Nobody makes you grow beets under such conditions. You can grow different crops with less trouble." Our growers understand all the dark sides of beet growing in our conditions with a four dollar per ton price, but agricultural conditions are stronger than all their estimates. Our fair farms cannot exist without crop rotations, and crop rotations could not exist without beets. Hard work we have raising the sugar beet, but our soil needs regular cultivation and beets give the highest grade of cultivation. Just think of about 7 or 8 cultivators of different kinds on a beet field during the growing season. Those cultivations refine our soil, enrich it with different chemical elements, expose it to the reaction of light and warmth, destroy weeds and insects' nests. Generally on a beet field we have an excellent yield of oats or peas the next year following beets.

For growing good crops of sugar beets as well as any kind of grain one question is absolutely indisputable—that is fall plowing. Fall plowing is the foundation of every system of crop rotation, even for a 3-year rotation. Those farmers and farms who by some means cannot introduce that system ought to abandon their farms for others. We grow almost all our crops on fall-plowed land. What does fall plowing give us? Readiness of the field for sowing at the very beginning of spring when everybody is interested in putting seed in soil having enough moisture for growth and warmed by spring's sun. Fall plowing exposes the nourishing elements of the soil to the reaction of light and warmth, and also to different physical, climatic conditions during a large period of the year. Fall plowing is a powerful destroyer of weeds and the nests of many insects exposing weed seeds and insects' eggs to the destructive influence of autumn's and winter's frosts and snow. Our fields now are absolutely clean of weeds. We had webworms the last time in 1902 and the badly damaged fields were principally those with weeds. Insects begin their work from weedy spots which are the best protectors for moths and their eggs. About 20 years ago we had almost every year different bugs on our grain and it was always the weed-infested field which suffered, or else it was grain on spring-plowed ground, or in the neighborhood of a field left unplowed the previous fall.

After such experience we prize fall plowing and I think it is not we alone. Those farmers in Colorado who are doing this may witness that fall plowing fits not only Russia.

Owing to the hospitality of The Great Western Sugar Company, I had the opportunity to look over farming in northern Colorado. I saw last spring the destructive work of insects on the beet fields. Where weak from drought beets were destroyed by webworms. Before I had seen many fields dirty with weeds and planted late where late planting after spring plowing could not catch enough moisture for the seed to grow. I saw the army of fieldmen of the Sugar Company flying over a large territory to help farmers fight insects and wild oats, and I saw many farmers working hard day and night with sprayers and Paris green or with hoes cutting wild oats. How many dollars were spent and how much energy was lost? What would have become of all the infested fields if the farmers had not back of them the skillful organization, the big resources and the ever-ready help of The Great Western Sugar Company? We never had such help and had to protect ourselves by early fall plowing. Why not save money and work and try this system on the fertile soil of Colorado?

With the many difficulties and hard hand work we grow sugar beets in southwestern Russia. From our sugar companies we don't get either such good prices for our beets or such big help—scientific and material—but we have always been glad to live in a beet-growing country. Owing to the sugar beet we built an intensive agricultural system on our land. We have systems of crop rotations which have made southwestern Russia the very best producing section in Europe. In the sugar industry we have a rural industry which creates common welfare in the surrounding communities. We have comparatively good roads and more railways than other parts of our vast country. On the by-products we raise good cattle and support our draft animals—horses and oxen. And last, enormous numbers of our peasantry have almost permanent work around the sugar factories.

By your sugar industry you have made certain the development of your country. With your methods of work, and with your numerous machines you have laid a solid foundation of the future welfare of your farming. This farming will flourish and will make your beautiful Colorado flourish. You live and work in a new, rich country and use all the improvements of your immense industry. With you, and always ready to meet your just needs, you have the big organization—the Sugar Company. The Great Western Sugar Co. has all kinds of experience, science and resources which are always used to help you. The energetic and progressive people of this company are interested in your successful farming as much as in the flourishing of the company. In your co-operation, co-operation built on mutual, sound interests is the foundation of a great future for your country.

These pictures were taken on the farm of C. L. Hover, west of Longmont. He stacked his grain about August 10 and the next week
started fall plowing.

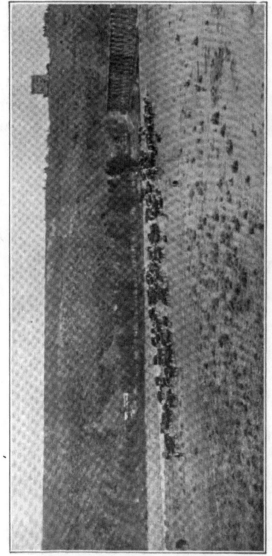

The ranges in Wyoming are very poor, owing to the drowth, and many stock raisers have had to ship elsewhere. Picture shows a bunch of cattle at Cody, Wyoming, waiting for cars for shipment.

424

Beets grown on farm of J. R. Chase, Lovell, Wyoming, season 1912. These beets were irrigated up April 25th, and thinning finished May 30th. These beets compare very favorably with any grown in Northern Colorado.

Government Irrigation Canal Supplying Powell Project In Northern Wyoming.

Some After the War Problems in Agriculture*

G. F. Warren
Cornell University

(Reprinted from Journal of Farm Economics, Vol. I, No. 1, June, 1919).

THE after-the-war problems in agriculture are not strikingly different in kind from the pre-war problems. But the war has made the problems more acute. We are now going through a period of reconstruction of ideas. All that is, is questioned. New theories on every subject receive a ready hearing. The world is in flux. That which is done may last for years. That which is not done may not be accomplished in years. If farmers do not now plan ahead, they may find that plans have been made as plans in the past so often have been made on the assumption that the problems of mankind begin at the city terminal of the railroad.

The Fundamental Rural Problem.—The fundamental problem in agriculture is to make and keep conditions of farm life such that a fair proportion of the intelligent and able citizens of the nation will continue to live on farms. Farm families are larger than city families. It, therefore, follows that whatever the farm population is, the nation will become. The strongest safeguard that the nation can have is an independent, foreward-looking and self-respecting farm population.

Methods of Meeting the Problem.—There are two theories as to the best way to solve the farm problem. One method is to search the world for persons who will be content with farm conditions as they are. This method has many powerful advocates. Some would bring in Chinese. Considerable agitation for this procedure is constantly going on. Others would bring in the backward races of Europe and Asia to work our farms—peoples so backward that to them our worst farm conditions would seem like luxury. The same idea often takes the form of complaint against the desire of the American farmer to share in the American standard of living. The conclusion is reached that the farmer should be replaced by a peasant family whose housekeeping is so simple that all members of the family work in the fields and whose desires for education are so slight that the children are kept out of school to work.

---

*Paper read before the American Farm Management Association at Baltimore, Md., January 9, 1919.

It is no merit in a peasant that he can pay for a farm as quickly or possibly more quickly than an American farmer can, when the latter keeps his children in school and allows his wife to devote a considerable portion of her time to caring for the home and children, rather than work full time at farm labor.

These contradictory ideas are not always thus boldly stated, but in practically every discussion of farm conditions each of these two points of view has its spokesman. Not infrequently the most plausible speaker advocates the wrong solution. Shall we make farm conditions such as to keep intelligence on the farm, or search the world for a civilization so backward that it will be satisfied with conditions as they are?

Movement from Farms to Cities.—In the past generation the conditions of living in cities have been greatly improved. It is evident, therefore, that unless corresponding improvements are made in farm conditions the intelligent portion of the farm population will be more strongly drawn to the cities than ever before. Let us see what these improvements are. Some of the more important changes may be classed under the headings of education, health, recreation.

The most powerful force that leads persons to leave farms is the expectation of greater remuneration. The majority of persons who go from the farm to the city go at one of three periods in their life; when the children must enter high school, when the farmer wishes to retire, or when young men and young women are old enough to start work for themselves.

The desire to allow the children to have high school privileges is one of the important factors that leads farm families to go to town. Not only is this one of the strongest factors leading families to go to cities, but it selects the very best type of citizens, that is the kind who are willing to make the most sacrifice for the benefit of their children. The remedy is obviously to bring high school facilities nearer to the farm.

Desire to live in a house that has a bath room, heat and electric light is an important factor in many cases. The remedy is obviously to make farming profitable enough so that farmers can afford furnace heat and bath rooms, and then develop a sentiment that will spend the money for a bath room, even though it may not add to the selling value of the farm.

The desire to be able to obtain medical service is another powerful factor leading middle-aged farmers to take their families to town. The remedy is to have better medical service in the country.

The desire for recreation is not one of the major factors in leading farm families to go to town, but does play a considerable part in the movement of young men and young women to the city. But the strongest force leading young men and young women to go to the cities is higher wages.

The only large demand for young women on the farms is to do housework or become wives. Many of those who do not marry farmers as soon as they are mature, seek employment in cities. The remedy is obviously to bring work to the country.

The desire for adventure that is present in every normal boy and girl may be satisfied in many ways without leaving the farm. Attendance at good vocational high schools tends to satisfy this desire by opening up the problems involved in man's attempt to conquer nature. The games and the social advantages of the high school also help.

The Increasing Cost of Living.—The high cost of living in each decade promises to become a more difficult question. A correct understanding of the problem is, therefore, of more than passing importance. We have, doubtless, passed the point of maximum food production per hour of human labor. New inventions help, but in spite of them, every additional bushel is now a more expensive bushel. A machine that saves labor on the farm does not save as much human time as is often assumed, for someone must make the machine. Food is becoming fundamentally more expensive to produce in terms of human effort, because poorer land must be used and because on the good land, production has reached the point of diminishing returns. If it were not necessary to increase the amount of food, inventions would reduce the amount of human effort required in food production. But the demand for more food calls for the use of land that must be reclaimed at great expense, and calls for more intensive methods on land now in use. It is of course possible, and perhaps probable, that improvements in manufacturing will take place so fast as to more than offset the increasing cost of food so that general well being may continue to be improved. But food is almost certain to continue to call for a larger share of the workers' income, if the population of the world continues to increase as it has in the past fifty years. There are no more Iowas waiting for the plow.

One of the great underlying factors in the present world conflict is the effort to place on someone the blame for the pressure of population on food supply. We can no longer obtain the former supply of food with the same effort. Not knowing that this is due to the ratio of population to natural resources, each class believes that it is not receiving just treatment. The industrially-minded believe that farmers are at fault, labor blames capital, farmers blame middlemen, consumers blame prices, nations blame each other.

The past generation was the golden age for manufacturers. It was the age of cheap food. We were harvesting nature's crop of lumber, and were skimming the fertility accumulated by ages of nature's processes. Now we must reclaim some southern soils where the hasty exploitation has caused erosion so serious that nature unaided could not remedy it in ages. We must get the alkali out of land on which our first dash of irrigation gave wonderful crops but serious consequences. We must fertilize soils that were at first productive, but that were not exceptionally rich in plant foods.

The capitalist, the laborer, and the city consumer agree on at least one thing. They are all unable to understand why the cheap food does not continue. They are willing to import peasant farmers, to entice soldiers to farms, to fix prices, to prohibit the killing of heifer calves, to do almost anything except the one inevitable thing, that is, pay more for farm produce than was formerly paid. Our rapidly multiplying population, the hordes that have been coming from Europe and the rather sudden running out of the free fertile lands, coming at the same time that labor organizations are demanding shorter hours and more of the comforts of life, make the problems of the near future acute. Add to all this the complications involved in deflation with the many injustices that follow contraction in the currency, and still further add the epidemic of mediaeval ideas that is spreading over the world, and we can well see the necessity of clear thinking. I believe that the American can solve the problem, if it is solved in an American manner, but if the German and Russian philosophies that are spreading over all the world are accepted, I am fearful of the future.

Each year when the rainfall is short the food problem is likely to be more critical than ever before. If the time comes when such a year is accompanied by unemployment, conditions will indeed be serious. In such years we may expect to see efforts to make food abundant and cheap by legislation. Such efforts have their natural reaction in desires of producers to have legislative protection in years of over-production and low prices. Both kinds of legislation are very dangerous.

There is grave danger that the present antagonisms between city and country will grow. There is at present no means of informing the consumer as to the farmer's point of view. The city newspaper is read by the farmer so that he learns the consumer's view, but as yet there is no effective means of giving the consumer the farmer's point of view. Nothing is so conducive to antagonism as lack of knowldege.

Nor does the farm point of view receive adequate expression in legislative halls or on the many bureaucratic commissions to which we are delegating the powers of government. Nearly everyone who has money enough to buy a home outside the cities is called a farmer, or calls himself one when he desires to discuss farm problems. Mr. Hoover is said to have remarked, "Who does represent farmers, and what do farmers really want?" The time has come when those who assume to speak for farmers should be representatives of farmers' organizations. Every such person should be an American first, but it is not enough to be an American when one deals with technical matters. He should also know technical details.

This is the age of organization. Class groups of all kinds are endeavoring to obtain for themselves more than their normal share of the good things of life. Nearly always the attempt is made to obtain the desired results by some form of monopolistic control. So far the producers of food have been about the only unorganized persons. Because of the actions of other groups, farmers are being compelled to organize. I am sorry to see organization come about in this way. I believe in organization, but am sorry that it has to come about for protection. I do not consider the real organization of society to be a collection of competing organizations.

The farmer does not ask an eight-hour day, he does not even ask a ten-hour day. But, except in emergencies, he is certainly entitled to expect to be able to make a living from ten hours of work. He does not ask that child labor be abolished on farms, but should ask that farm children be not kept out of school to do farm work.

At the risk of appearing to add a few planks to the agricultural platform, I will mention in detail a few of the farm problems as I see them. I have omitted some of the most important problems, but certainly all that I mention are worthy of consideration.

Unemployment

During the war, food prices have been controlled to some extent. This control has had some influence in holding down production. Probably it had delayed to a limited extent the time for highest prices for food. As wages come down there may be more critical food problems than we have yet met. The greatest danger is that a short crop year such as occasionally comes may come at the same time as a period of unemployment.

The farmer is as much injured by general unemployment as is the city. During the period of reconstruction and deflation of the currency, it is highly probable that periods of panic and unemployment will come. Plans should be made for great public works and held in abeyance to be developed whenever unemployment occurs. This will be the cheapest way to get such work done and will at the same time help to stabilize conditions. Such work in the past has usually been done in periods of high prices rather than in periods of unemployment. There are many improvements that the Federal Government, states, and local governments should make. Roads, new school buildings, and hospitals are among these. The Federal Government should construct good permanent buildings in Washington to avoid paying the enormous rents that it has for years paid. The war has shown the need for a few national highways for moving freight and passengers. One such road should run north and south along the Atlantic coast, one along the Pacific, and one down the Mississippi valley. One should cross the northern part of the continent and one the southern part. Ordinary state roads are wholly inadequate to handle the traffic that the government has put on the section of road from Chicago to New York. These trunk lines are spectacular, but the greatest need is for better roads from farms to the railroads. Both are needed . States and counties should be continually building roads but they ought to prepare comprehensive plans and push the work when it can be done cheaply, that is, at periods when there is unemployment.

Protection

The first duty of a state is to protect its citizens. The city dweller calls the police at the first sign of trouble. In most states, the farmer must turn to his shotgun, just as his forefathers turned to the rifle. In Canada and in some of our states mounted police now bring protection to the farmer and his property. This protection is particularly needed in the states that have a dense population, and in the South, but every state should have police protection in the country. The common practice in small towns of avoiding expense by allowing petty criminals and hoboes to go free if they will leave town often results in sending them to the country.

Education

Free education in every subject from the primary grades through the university is the only sound basis for democratic citizenship. This means that provision should be made for allowing boys and girls to attend any public school without having to pay any tuition. I believe that free high schools are now available for all children living in cities, but a very large number of the farm children have to pay tuition. Even where tuition is free the farm children often have to leave home and pay board so that education for them is very expensive. The number of small high schools should be increased so that education may be made available to all. State aid should help small public schools and small high schools to such an extent that equal school privileges can be had in all parts of the state by substantially equal local taxation. Education is a state function. If the state desires to have persons live on farms it should see to it that educational opportunities are not thereby denied. The farm problem will not be solved so long as the accident of being born on a farm denies so many children the privilege of attending high school.

Laws should be so framed as to have state aid follow the boy or girl to any public high school in the state, and every high school that receives any aid from the state should be open free to every boy or girl in the state, or else all state aid for that school should stop.

Not only should there be free common schools and free high

schools but college and university education should likewise be free. In some of the older states, all of the original arguments against free common schools will be brought to bear against this proposition and each of these arguments will be refuted by the same logic that made the free common school triumph. I will not attempt to enumerate these arguments here. Many states have for years furnished free education to all. It is no new experiment, but a sound principle of democracy that must be accepted. Now is the time to demand it. Any ablebodied boy or girl who does not have to help to support others, can make a living while attending college. But to add $100 to $150 a year for tuition makes such a course prohibitive to many.

Labor

About three-fourths of the farm labor is done by the farmer and members of his family. When farm wages are high the farmer and his family receive good pay for their work, when wages are low they receive poor pay.

The individual farmer sometimes thinks that if he can get cheap labor it will help him, not realizing that when all get the same kind of labor, the labor that he and his family do must compete with the cheap labor that he has helped to introduce. Except in the South there is no permanent hired-man class in America. The hired men are in general, brothers and sons of farmers. Whenever any other type of labor is largely introduced into a farming community, social conditions become very bad. Population is so sparse in all farm regions that there are not enough people to keep up good social institutions unless all the people are of one race so that there are no impassable social barriers. Every farm community should do all in its power to prevent the introduction of any kind of laborers who do not promise to be assimilated.

The employer should always have a serious labor problem in a democracy, for a labor problem usually means that labor receives so large a part of what it produces that great skill is necessary in employing it in such a manner as to be able to pay the common wage.

Land Ownership

The American ideal in handling land is to have the operator be the owner. Most of the farming in America is done on this basis. Normally the farmer who is too old to continue farming rents his farm to his son, or son-in-law, or to some other young man who as a hired man in the community has established a reputation for honesty and thrift. Later the farm is sold, usually to a tenant. The average amount of time spent as a tenant in New York is about ten years, and the average tenant becomes an owner at about thirty-five years. This means that nearly half of the tenants are not able to become owners until they are older than 35 years. Conditions are approximately similar in other regions where not more than one-third of the farmers are tenants. It would be desirable to have credit systems so perfected that the average age at which ownership is acquired could be reduced to thirty years so that more of the years when the farmer is at his best physically could be devoted to making a home, and doing the innumerable things that need to be done on farms.

In some parts of the United States land tenure has become a serious problem. There are several ways of meeting the situation. Perfection of credit systems is one way. Other proposed remedies have to do with laws concerning land ownership.

There are fundamental reasons why individual ownership of agricultural land is the only sound basis for agricultural development. Theories as to single tax, and nationalization of land are widely promulgated. Such theories are always city-made. They fail to distinguish between city building lots and farm soils. They are able to see buildings as an improvement made by the owner, but do not realize that a farm soil is also made or destroyed by the owner. Any farmer in an old community knows that a soil that is worth $100 per acre is little more than pay for the drains put in it, the stones picked off, the fences put up, the weeds eliminated, and the residual manure, and fertilizer applied. English laws concerning tenant rights recognize roads, stream control, tile drains, fences, buildings, orchards, grass seeding, permanent pasture, residual manure and residual fertilizer as among the improvements for which the tenant is entitled to compensation, that is, values added by the operator. The true farmer watches and cares for his fields as he does his flocks. His fields are personal. He does not see the bacteria in the soil, but by indirect means he raises bacteria and earthworms as carefully as he husbands his flocks. A generation of farming in which the soil does not receive this personal regard is enough to ruin any but the best land.

But private ownership must not be abused. No farm land should be allowed to be held continuously for purposes of renting. Every community should include a limited number of tenant farms as aids to young men in getting started. But this number is fully supplied by farmers and widows of farmers who rent their farms for a few years after they are unable to continue direct operation, and before they are able to make a satisfactory sale. It often takes five to ten years to sell a farm at a satisfactory price. The landlord who has been a farmer often contributes very materially to the success of the farm and the tenant because he knows how to farm. Just how long the owner should be allowed to run a farm as a tenant farm is a question, but certainly this period should not exceed twenty-five years.

Large tracts of land that are too large to be operated to public advantage should be broken up into farms of normal size and sold to the operators.

As important as these prohibitions, is positive legislation that will bring together the persons who have money to lend and the young farmer who wishes to establish a farm home.

The Federal Land Bank promises to be of much help. The law may need to be revised, but it should not be revised in any manner that will increase the amounts loaned per farm. The land bank loan should represent an exceedingly conservative loan in order that it may carry the lowest interest rate.

Certainly legislation is needed that will do for agriculture what the Federal Reserve system is doing for commercial industries in furnishing short-time credit. Possibly this can be accomplished by modifying the Federal Reserve law, or by supplemental legislation. Such credit in agriculture must recognize any period up to one year as the normal basis for a short-time loan.

Mere lowering of interest rates is not the primary factor in credit. Interest rates and land values bear considerable relationship to each other. The time allowed and methods of payment are quite as important as interest rates.

Food Distribution

I am not in sympathy with blind attacks on middlemen any more than with blind attacks on farmers. The individual middleman is often controlled by general circumstances over which, as an individual, he has little control. But the feeding of great cities is a new problem. It is not to be expected that our present methods are the best ones than can be devised. In some cases radical changes must be made.

A few great handlers of food are obtaining control of more and more kinds of foods. The public will not long tolerate any "hold-ups" by persons located at strategic points on the road from producer to consumer. Some of the bitterest contests of the near future promise to be over food distribution.

The agricultural colleges should begin serious study of the problems of food distribution. Such studies are very different from fertilizer tests or feeding experiments, and will require a great broadening in the outlook of the agricultural colleges.

Co-operation

Many public agencies have been advocating that farmers co-operate in the sale of their produce. There are many instances of successful co-operation. However, in those cases were the farmers' organizations must deal with large corporations that approach monopolistic proportions the road to co-operation is by no means free from obstructions. In some states one branch of the government has been preaching co-operation and has been organizing co-operative societies while another branch of the state government has been doing the "follow-up" work by trying to put the officers of the co-operative associations in jail because they co-operated.

The farmer is a laborer who owns his own tools; he is also a small business man. His products are usually more the product of his own toil than they are the product of capital, or hired labor. The principle of collective bargaining must be accepted for both classes of labor. This does not mean that an unscrupulous labor body that happens to hold the key to public welfare should be allowed unlimited action, nor should an unscrupulous group of farmers be allowed unlimited action. We do not allow unlimited action to two men who are making a horse trade. But, the principle of collective bargaining should be recognized.

Equality of Opportunity

A democracy must allow to every individual complete freedom to enter any useful work that he may choose. The right to enter any occupation and make of one's talents the fullest possible use is fundamental. So long as complete freedom in choice of occupation exists, it is utter folly to attempt to entice persons into any particular occupation. Those who are added to the industry by special endeavor only force others out of the industry. The government should furnish education that will help each person to decide on his occupation, and should furnish technical training to help in preparing for the chosen work, but the only democratic way to maintain the proper proportion of workers in each industry is to have the rewards for a given ability and effort the same in one industry as in another.

Substitution

The wooden nutmeg was looked upon as a great achievement in the era of substitution. But progress since that day has been so rapid that the wooden nutmeg looks as primitive as a high-wheeled bicycle.

With mounting costs of living, the desire to use everything is commendable, so long as the article is wholesome and is sold honestly. But substitutes are not noted for their honesty. Oleomargerine would have few enemies if the groceryman sold it as oleo and if the bill of fare called for bread and oleo instead of bread and butter. The manufacturer wants to color it yellow for the sole purpose of deception. It is not that yellow is an especially favored color. Meat substitutes are not colored yellow, and if some one invents a substitute for an apple he will want to color it red, not yellow. When cottonseed oil is used to make imitation butter, it is colored yellow, but when the same oil is to pass for lard, white is perfectly satisfactory. Oleo now has its own troubles, for still cheaper oils have been found and are sold as oleo when they are not. We now have imitation eggs, and imitation condensed milk, and imitation cow feeds. Aside from the deception which in itself is injurious, there are other serious effects. Children that need animal food are fed on vegetable fats that will not make them grow, and the legitimate industry is seriously hampered by having to carry numerous parasites. Animal foods are always more expensive than vegetable foods, but they are necessary. The fight to compel all products to be sold for what they are has not been won.

It is now time for all producers and consumers to take a definite stand that no injurious substances shall be allowed to be sold, and that every product shall reach the ultimate consumer for what it is.

Dissemination of Manufacturing

The exceedingly cheap food in America has to a considerable extent been responsible for the congestion of our cities. Food was so cheap that the industrial worker did not care for a garden. He could live on the tenth story and have food brought to him. The high prices of food promise to favor the manufacturing plant that is located in a region where the workers can have small areas of ground for gardens. Here women and children can work, and the laborer can work out doors after hours. A limited amount of gardening after a day in the factory is the best kind of diversion. It is time to stop building vertically and begin to build horizontally. This movement has already begun.

Public agencies such as the agricultural colleges and department of agriculture should study the problems of farm economics as they have in the past studied the problems of production. Such studies will require good judgment and tact, but the tact should not go to the extent of failing to tell the truth.

These are but a few of the problems that we must now face. Farmers should be fully organized so that they may see to it that these and other national problems are solved in the American manner, rather than be solved by imported ideas brought over by backward nations.

NEW WHEAT DISEASES PROBABLY ERADICATED

Indications are that the two dreaded foreign foes of wheat, flag smut and take-all, will not become wide-spread in the United States. The United States Department of Agriculture announces that the two states where these diseases appeared, Indiana and Illinois, have taken steps that will prevent the spread of the disease from the infected fields and that should wipe out in a few years the infection in fields where it exists.

Indiana officials came to the recent hearing in Washington with adequate safeguards already placed. Shortly after the hearing, Illinois established similar safeguards. All the infected wheat in both states is under control and will be disinfected before any use whatever is made of it. All straw and stubble are to be burned, thrashing machines are to be thoroughly disinfected, and no wheat is to be grown in infected areas for several years.—Agricultural Review.

Banish Scrub Sires from United States

With confidence that better live stock will bring satisfaction and many benefits to its owners, I invite the concerted action of farmers, stockmen, and others in banishing scrub sires from the United States. The widespread use of inferior male breeding animals has been for many years a cause of low production per animal and of needlessly poor quality.

The continuance of such conditions is uneconomic and unnecessary. The direct and practical means of improvement is to use breeding animals, especially sires, which are true representatives of breeds developed for a definite, useful purpose.

I am confident, too, that the public, knowing production to be performed with maximum efficiency, will look upon stock raising with increased respect and understanding. In a few localities noteworthy improvement in special lines has been taking place through individual and community efforts. Let us now hasten such improvement wherever live stock is kept in the United States.

J. R. MOHLER,
Chief, Bureau of Animal Industry.

An Enterprise Study in Four Sugar Beet Areas

L. A. Moorhouse

URING the summer of 1910 farm practice studies were begun in the beet-growing districts of Colorado. In a few limited areas the farm operators had encountered some difficulty in securing satisfactory yields. It was rthought that a review of the practices on several of the more successful farms would provide some suggestions that could be used to advantage in helping the men who were getting low yields. Observations were also made on farms where the returns from the beet crop were comparatively low. This work was continued for two years and several preliminary reports were prepared outlining and describing the conditions as determined at that time. Practically no new studies of this character were made in 1913 and 1914, but the project was placed on an active basis in 1915. A systematic effort was made throughout that year to obtain a reasonable number of enterprise records in Colorado and Utah. In 1916 these studies were repeated in Colorado and Utah and additional records were obtained in Michigan, Ohio, Idaho and California. The work was continued in California in 1917. It may also be stated that supplementary data concerning the labor requirement of crops other than beets have been taken in two of these regions. We also have complete farm survey records for limited areas in three regions. This paper will be confined, however, to the single enterprise, sugar beet production.

It is my purpose in presenting this paper to point out some of the practical features that have come to our attention as a result of this survey. Briefly I wish to make an appraisal of the enterprise survey. In the first place, a study of this character reveals the actual field practice employed in handling the crop. I am speaking now of the sugar beet in particular. Such facts are of value and serve in a suggestive way to the man who is operating in a region that possesses features somewhat different from others districts where this crop happens to be one of the leading staples. In taking the records which served as the basis for the few reports* already issued, we had

---

*See Dept. Bul. No. 693, Farm Practice in Growing Sugar Beets for Three Districts in Utah and Idaho; Dept. Bul. No. 726, Farm Practice in Growing Sugar Beets for Three Districts in Colorado; Dept. Bul. No. 735, Farm Practice in Growing Sugar Beets in the BillingsRegion of Montana; Dept. Bul. No. 748, Farm Practice in Growing Beets in Mich. and Ohio; Dept. Bul. No. 760, Farm Practice in Growing Beets in Three California Districts; Farmers Bul. No. 1042, Saving Man Labor in Sugar Beet Fields.

Fig. 1.—Different Types of Cutivators Used in the Intertillage of Sugar beets.

the privilege of discussing with beet growers in the humid belt some of the problems that obtain in the irrigated valleys of Colorado, Utah, and Idaho. Likewise we found the irrigated farmer interested in the problems of the Michigan beet grower. While the conditions in these two areas differ widely, at the same time it is possible to find definite practices in one region that have something of a lesson in them for the man who is operating in an entirely different area. This in turn will lead to the adoption of new and improved methods, and the enterprise can be placed upon a more efficient working basis. Special illustrations could be cited to show how the practice of one area has influenced in a minor degree the practice in an adjacent area or region. This particular industry was developed several years ago in Utah. Colorado introduced the crop at a later period. Men who were familiar with the industry in Utah were brought over into Colorado to aid in organizing the work, and to give advice with reference to cultural practices. Some of the features that proved to be successful in the Salt Lake valley were subsequently transferred to the Colorado fields.

Such facts as are brought out in a study of field practice are of value also from the historical standpoint. By reviewing the records which have been taken and consulting the reports which have been issued, we can secure a correct picture of the practices that were followed in these regions during the years 1915, 1916 and 1917. If similar surveys can be made in these same regions about ten years hence, it will be possible to determine the changes that have taken place within the decade. The points I have in mind relate not only to manurial and fertilizing practices, but they also include such operations as plowing, harrowing, disking, planting, harvesting, etc. In the Garland area of Utah it was the practice on some farms in 1915 to allow the alfalfa to remain intact about five years. The field was then broken and planted to grain. Beets came after grain for five years or more in succession. At the end of this period grain was sown on the beet land and the field was reseeded once more to alfalfa. It will certainly be of interest to know ten years hence if this cropping system has been changed to any appreciable extent. In the Greeley area of Colorado it was customary to leave the alfalfa down from three to four years. If the operator gave any attention to the production of potatoes, this crop was planted as soon as the alfalfa was broken up. Sugar beets followed the potatoes for two years or in some instances three years. Peas or beans came after beets, then the field was sown to grain and reseeded to alfalfa at the same time. Generally speaking, farm manure was applied to beet land. The ground was plowed quite largely in the early spring. The two-way plow was used almost exclusively in this area, and a large number of the operators employed a crew of one man and four horses in

441

LABOR REQUIREMENTS IN PRODUCING SUGAR BEETS

1085 Records

District	Aver-age yield (tons)	Acres grown (total)	Farmers' labor Hand Opera-tions Hours	Machine Opera-tions	Horse Opera-tions	Contract or hand operat. Hand Equiv-alent tions $	Ex-pended hours	Total hand operat. hours	Man Equiv. labor hours	Total hours per acre Man labor	Horse labor
Garland, Utah	14.85	1461	36.7		21.2	18.87	75.4	96.6	133.3	98.5	
Provo, Utah	14.96	855	58.8		46.4	5.90	25.6	72.0	130.3	117.14	
Idaho Falls, Ida.	15.62	735	34.2		16.0	17.29	69.2	85.8	119.4	79.5	
Greeley, Colorado	15.57	5028	48.5		6.3	17.26	69.1	75.4	123.3	104.5	
Ft. Morgan, Colo.	15.65	2456	45.3		18.7	13.52	54.1	72.8	118.1	105.0	
Rocky Ford, Colo.	12.99	2429	56.0		4.9	14.11	56.4	61.5	117.3	132.7	
Los Angeles, Calif.	14.52	7712	27.7		----	15.01	60.0	60.0	87.7	109.3	
Oxnard, Calif.	9.53	2811	20.2		----	14.02	59.3	59.3	79.5	111.5	
Salinas, Calif.	15.59	3616	35.7		----	18.07	75.5	75.5	101.2	124.3	
Caro, Michigan	9.72	2018	39.45		5.09	15.26	61.09	66.0	105.5	80.05	
Alma, Michigan	11.40	506	50.53		10.32	13.65	54.2	64.62	114.8	95.34	
Grand Rapids, Mich.	10.16	231	45.3		16.4	12.66	50.6	66.0	111.3	93.88	
Northwestern Ohio	15.17	1525	38.55		6.82	17.24	69.0	74.82	113.4	79.12	

doing this work. Very little disking was done of these fields. The land was leveled, and harrowed about three times with a spike-toothed implement; the field was rolled once, in some cases before and in other cases after planting, cultivated approximately four times, furrowed out once, and irrigated approximately three times. At the time these records were taken tractor power was used to a very limted extent on these farms. Just how far the field practice in handling these crops may change in a decade can be determined only by repeating the survey a few years hence.

An enterprise survey will indicate the most efficient as well as the least efficient methods of doing the various kinds of work directly connected with the production of the crop. One illustration will serve to emphasize this point. In the Michigan and Ohio areas, three types of cultivators were used. These types included one, two and four-row implements. The general summary tables indicated a marked difference in the labor cost of this operation, and further tabulations pointed out that it required 7.5 man-hours, and 7.05 horse-hours to cultivate an acre of land four times over with a one-row crew, using a two-row implement, whereas the labor requirement for the four-row implement utilizing a crew of one man and two horses was 3.8 man-hours and 7.7 horse-hours. The labor cost per acre in the former case was $2.27, whereas in the latter case this work was done at a cost of $1.54 per acre. Several additional practices could be cited to show that it is possible to secure considerable efficiency material from an enterprise study.

An investigation of this nature also furnishes basic data with reference to the labor requirements of the crop under study. It will also indicate the more stable factors that are useful in working out the approximate cost per acre or the cost per ton for any given period. In this connection, I wish to present a general summary table showing the labor requirements of the sugar beet as determined by this survey. (See opposite.)

I wish to direct attention to two or three important items that may be obtained by an examination of these figures. It should be explained, in the first place, that the cash expenditure for contract labor has been converted in this table to equivalent hours by dividing through at the rate of twenty-five cents per hour. The hours devoted to machine operation plus the hours of hand work done by the farmer or his family plus equivalent hours will give the total man labor per acre. It will be seen that Provo farmers performed a large part of the hand work. This is a region of small farms. In three California districts the hand labor was all done on a contract basis.

In the Billings area, Montana, it was found that 41.7 hours of man labor were required for the machine operations. The horse labor amounted to 94.1 hours per acre. The contract labor included a total

444

Fig. 2.—Variations in Crew Size for Spring Tooth Harrowing.
of 93.2 hours per acre, thus making a total of 134.9 hours of man
labor per acre in this region. This is comparable with the Garland
region and is similar to the Greeley area. Minnesota Bulletin No. 154
shows "that the labor requirement in producing sugar beets in that
state amounted to 155.4 man-hours and 154.7 horse-hours per acre.
The author of this bulletin in commenting on these figures states
"that according to table 4 of this same bulletin, the professional beet
worker performed the bunching and thinning, the hoeing, the pulling
and topping in 19.5 less hours than the farmer, thereby reducing the
total man hours per acre to 136. When professional beet workers are
employed the farmer performs the operations except those mentioned,
and the labor of the farmer amounts to but 51.5 hours, indicating that
61 per cent of the labor is concerned with the three important hand
operations." The figures for the Minnesota work are somewhat
higher than those indicated for the Caro, Michigan, and northwestern

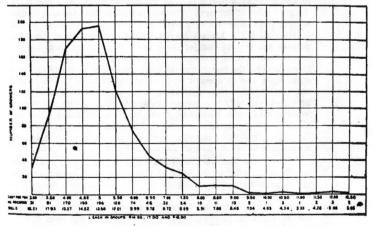

**Fig. 3.—Frequency Curve Showing Distribution of 1025 Operators and
Relation Between Cost per Ton and Yield per Acre.**

Ohio districts, but are quite similar to our results for the Alma and
Grand Rapids areas of Michigan.

These studies emphasize the importance of securing good yields
in order to lower the cost per unit of production. The comparisons
and observations which have been made in these districts have shown
why some farmers are able to get returns of 18 to 20 tons while
neighboring growers secured from 9 to 10 tons per acre. The accom-

panying frequency curve (Fig. 3) indicates the relationship that exists between yield per acre and cost per ton in producing the crop on 1025 farms. The lowest individual cost was $2.54 per ton, while the highest was $18.16 per ton. The average cost for all districts was $4.70 per ton.

Seventy-three and one-half per cent of the growers had a cost of $4.00 to $6.00 per ton, while the three groups from $4.00 to $5.00 per ton represented fifty-four and one-half per cent of the growers interviewed.

This project has been carried on on a co-operative basis with the Office of Sugar Plant Investigations of the Bureau of Plant Industry. I requested Dr. C. O. Townsend to give me a statement concerning the value of this work from his point of view. He writes as follows:

"The sugar beet grower, in order to accomplish the highest results, must have clearly in mind the various operations necessary in the production of this crop. These bulletins set forth these operations in proper sequence in a very striking manner. Furthermore, in order to accomplish the highest results the beet grower must know the purpose and cost of each of these operations. These facts are also brought out in these bulletins. These studies also impress upon the farmer's mind the importance of these operations in connection with sugar beets and as a result he gives closer attention to the purpose and cost of the various operations in the production of other crops. This leads naturally to a comparison of crops grown with and without rotation with sugar beets, and in this way he establishes in his own mind by actual practice the value and importance of sugar beet culture. He gives closer attention to the purpose of each operation and seeks for a better method of accomplishing the desired result.

"I feel that this work is of real value to the farmer and is very helpful in carrying on the work of this office along agronomic lines. These bulletins, bringing out as they do the methods followed in sugar beet culture in each of the sections studied, suggest to the workers in this office the lines of investigation and study that need special attention. For this reason the work that has been done by the Office of Farm Management in regard to sugar beets is of the greatest importance to this office. I hope it will be possible to repeat these surveys in practically the same localities, covering the five general sugar beet areas, and to extend this work somewhat in order to cover a larger area and furnish additional data."

Conserve moisture by discing before plowing

The Silo---The Lighthouse of the Farm

George Keppel

 ILOS gave the dairymen of Minnesota some light on economical feeding last year. In seventeen cow-testing associations 3,108 cows were fed silage, and 2,385 received no silage. The silage fed cows produced an average of 5,549 pounds of milk a head a year, while those receiving no silage averaged 4,274 pounds a head. That is a difference of 1,275 pounds in favor of silage. The silage fed cows averaged 215 pounds of butter fat a head for the year; those receiving no silage averaged only 164 pounds a head—fifty-one pounds less than the silage-fed cows.

The cost of feed consumed by the silage-fed cows was $31.52 a cow; the cost of feed for those not fed silage was $20.81, or $10.71 less than for silage fed cows. While the total feed cost for silage fed cows was greater, the net returns above feed cost also were greater.

The net returns for silage fed cows were $42.44 a head, and for those not fed silage $33.31. That is a difference of $9.13 a head in favor of silage. In a ten-cow herd this means an added profit of $91.30 a year from feeding silage. At that rate it would take but a few years for the silo to pay for itself. A return of $9.13 a cow is good interest on the money required to build a silo.

Dairying is not the only class of farming in which the silo is the lighthouse of the farm. A silo will pay on a general farm where ten head of cattle, or the equivalent of sheep and cattle, are kept. Silage is not used so extensively for horses as for cows and sheep, although some farmers give reports on feeding silage of first-class quality to horses and mules. Occasionally a swine raiser uses silage for brood sows.

Many who have a small heard of cows figure they do not keep enough to warrant the building and filling of a silo. This is often a mistake, for with the small herd the overhead cost of care and management is often higher than it is with a large herd and the need for economy is still greater.

Advantages of a Silo:

1. It saves practically all of the food values in the corn crop, especially in case of early frost.

2. It saves the food values of the corn plant in a palatable form—cattle relish silage.

3. It gives you a feed in midwinter that has all the features of a June pasture.

4. It insures a saving of time and labor in feeding.

447

5. It will store the corn crop, grain, fodder, and all in the smallest possible space.

6. It allows the use of larger type corn; therefore more corn can be grown per acre.

7. It provides feed that increases flow of milk in winter and lowers cost of production.

8. Along with good clover or leguminous hay, silage requires very little additional feed for milk, and a minimum of grain for beef.

It is difficult to determine accurately the saving made by the silo when the cost of the ration is considered, for so many conditions enter in which make a variation in the result; but it is safe to assume that from $10 to $25 can be saved in the cost of feeding a cow one year by the use of the silo.

It is also safe to say that with silage the cost of butter-fat can be reduced from six to twelve cents a pound; the cost of producing 100 pounds of milk from thirty to fifty cents; 100 pounds of beef from $1 to $2; 100 pounds of mutton from seventy-five cents to $1.50. By properly arranging the silo and the barn it is possible for one man to feed forty cows in thirty minutes and the work can be done with ease and without the use of a horse or wagon. Eight times more feed can be kept in a silo than in a haymow.

A silo will make your acres earn more, your stock earn more, your work easier, your storage greater, the production of your cows higher, your farm richer; and with these advantages you will derive more pleasure and profit from your labor.

Crops for the silo.—Corn is king of silage plants on account of its high yield and universal use in the stock growing country.

Kafir-corn makes excellent silage for all kinds of live stock, and its drought-resisting qualities make it a dependable crop for western and southwestern states. Milo maize, sorghum, fererita and Sudan grass are also very valuable as silage plants, being drought-resisting and yielding a high tonnage per acre.

Root crops, such as sugar beets, mangels, rutabagas and turnips, can also be put into the silo with good success. Beet pulp from sugar factories has been used successfully as silage.

When to cut silage corn.—By all means avoid cutting silage corn too early. Silage from immature corn turns sour, is less palatable than when properly made, and has lower feeding value.

Best results are obtained by commencing to fill while the leaves of the upper part of the stalks are yet green, but not until the kernels are distinctly dented, in dent varieties, or considerably hardened in others.

When silage corn is so dry that the cut fodder does not feel moist as it is squeezed in the hand, water should be added at filling time. This can best be accomplished by turning a running hose into the blower. The amount of water to be used will depend on the rate of filling. Aim to make the cut fodder moist enough to pack down solidly.

A silo survey recently made in Grays Harbor County, Wash., indicates that a machine of relatively small size is the most economical to use in filling silos. The smaller machines require a smaller crew and less cash need be expended for help. One or two hours lost is not so serious as if a large crew were idle. Another advantage in using a small machine owned by three or four farmers in a community is the greater length of time allowed for settling, increasing the silo's capacity.

There are several kinds of silos, all of which are dependable. The important thing is not so much the kind of silo, as it is the way the silo is built. The chief things to consider in choosing which kind are cost and durability. One of the simplest forms of silos is the pit silo, constructed like an ordinary dug well.

Concrete silos have this advantage: They are very durable and will last for years, as will one of hollow clay tile. Cement stave silos are durable, but are not so stable as poured concrete. Brick silos will last a long time.

Wooden stave silos have this advantage: They are cheap, easily and quickly built. Wooden silos are more numerous than any other type. They last from eight to fifteen years if kept in repair. Metal silos give satisfaction.

Those intending to build their own silos should write to Chief of Division of Publications, Department of Agriculture, Washington, D. C., and ask for Farmers' Bulletin 855, which tells how to build home-made silos.　　　　　　　　　　　　—From The Farm Journal.

$1,020 FOR A HALF BUSHEL OF WHEAT

It is possible for the best half bushel of wheat displayed at the International Soil-Products Exposition this year, to win $1,020 in cash and trophy prizes. The least it can win will be $385.

Champion wheat growers throughout the United States and Canada enter into the most lively competition each year for these great prizes. Canadian growers have been winning for several years, but by the closest margins, over wheat grown in the United States. Seager Wheeler, of Rosthern, Saskatchewan, was the successful man last year, making the third winning of this great prize for Mr. Wheeler within five years.

Determined efforts will be made this year by many prominent wheat growers in Kansas, Missouri, Nebraska, Colorado, Oklahoma and other states to keep the trophies and cash prizes in this country this year.

The International Soil-Products Exposition is staged by the International Farm Congress. It will be held this year at Kansas City, Mo., September 24 to October 4.　　　　—Agricultural Review.

449

When and How to Silo Corn

C. H. Hinman
President Hinman Silo Co., Denver, Colo.

Should Be Mature.

Corn for the silo should be well matured. The corn plant, in common with all other plants of the same family, gains in food value all the time until its final maturity. The increase in food value is more rapid during the formation of the ears and filling of the kernels than in any preceding stage. Where silage is to be used for milk production it is not so important that the corn be fully matured as it is where the silage is to be used for fattening. For the latter purpose a variety of corn which will mature in this climate should be selected and this corn should be allowed to mature. When mature the lower leaves and some of the outside husks will be dry and the corn will be all dented.

Water May Be Needed.

In case early frost or extreme drouth has caused the corn to be abnormally dry, it is necessary to add water to the cut feed as it goes into the silo. This is usually done by the use of a hose running into the blower and one-half to one barrel of water per load should be used, according to the dryness of the fodder.

Cut Clean.

It is important that the fodder be cut clean, that is, that the knives be sharp and set close to the shear-bar so that all leaves may be cut. The general practice is to cut in one-half inch lengths. Our own preference is for one-fourth inch. We realize that this requires more time in filling, but with the fine cut it is not necessary to add water unless the corn is very dry, and the resulting silage is less sour and makes a more valuable feed.

Pack Thoroughly.

The packing of ensilage is very important. It assists both in quality and capacity. It is economy to use a jointed distributor, for with this all parts of the plant will be deposited together and at least one man in the silo can be dispensed with. The labor element alone will pay for a distributor in one season, but there is also a saving in that the distributor conveys the corn, leaves and stalks to any part of the silo desired and insures a better keeping and packing. Give special attention to the tramping at the outside edge next to the wall. See to it that all unnecessary obstacles on the inside of the walls are removed. The wall of the silo should be smooth and tight. Make the doors as tight as possible. As the process of filling advances towards the top, packing will be more necessary than at the bottom and extra help can be applied at this time. If the silo is refilled it should be done before the top layer has spoiled. The greatest settling period will take place in from thirty-six to forty-eight hours after filling. Two or three days settling will be ample, but longer will do no harm. With large filling equipment extra help should be used in tramping the silage.

Top Will Spoil.

It is difficult to preserve all of the corn in a silo. As a rule a few inches on top will spoil. This spoilage can be greatly reduced by putting on the top some very green corn or cane. A small patch of corn where ears have been removed will answer fully as well, and sunflower, weeds, or hay may also be used.

Much care should be used in packing the top. Tramp thoroughly over all surface and after it has gone through the sweat for a couple of days, when top material is spongy and warm, tramp again, giving care to pack thoroughly the edges. This will insure good keeping and prevent much loss at top. If silage is to be used immediately after filling, no spoiling is necessary for the silage can be taken out and fed during the process of fermentation. When spoiled silage is taken off the top it should be put where stock, especially horses, will not eat it.

Summary.

We can briefly sum up the subject as follows:

Cut ripest corn first.

Cut in small bundles.

Be sure to do a clean-cut job and cut fine.

Oil all running parts of cutter frequently and keep rollers adjusted to hold silage tightly.

See that cutter is given proper speed. This should be figured out before starting and proper pulley secured.

Tramp ensilage thoroughly.

Keep the knives sharp—the cutter bar sharp and keep the knives adjusted close up to the cutter bar.

Feed evenly—don't over-crowd the machine, but keep it full.

Plowing can be done more cheaply now than in the spring

Selecting a Beet Puller

T. A. Smith

AS beet harvest is drawing near again those who have beet pullers begin getting them in shape for the coming harvest, while others have to buy new ones. In driving over my territory I see a number of pullers standing in the fields where the last beets were pulled last fall. This does not look good, for we know there are but very few farmers but who have shed room for such a smal piece of machinery. After a puller has stood in the field for a year with the points rusting off, the price of it will soon be lost by the beets that are left in the field not pulled and broken off. This fall when the beets are all pulled do not forget to take the puller to the shed, for it is just as easily done as to unhitch from it, leave it in the field and drive the team to the barn.

All pullers do not work satisfactorily, in the different soils we have to contend with. This year we have growers who have never grown beets before, therefore they will have to buy pullers. The cost of pullers on the market makes it necessary that the utmost precaution be used in selecting the one most satisfactory.

The pullers made by the older implement manufacturers have in most cases proved quite satisfactory. However, there are some pullers that have been put on the market quite recently that are not adapted for use on all kinds of soil. There is one make of puller particularly, which when used on heavy soils leaves a large number of beets in the field. Any puller that runs more on the points the deeper it is run isn't nearly as efficient as the one that can be adjusted so that the heel (or back part of the points) can be lowered the same in proportion.

The old two-point walking pullers did far better work than any sulkey puller on the market today. If the old-style points were used on the sulkey pullers there would be less beets left standing in the fields after harvest.

The John Deere puller, in the writer's opinion, comes nearer meeting all soil conditions than any other sulkey puller. The beets are left standing where pulled and not bunched up as most pullers do. In this way they are not exposed to the hot sun early in the harvest, or to the frost in late harvest. Also all beets that have been pulled and not piled during the day will not be frosted as would be true with most pullers.

The P. & O. puller gave very good satisfaction last fall, what few I saw working. Also the Giddings; but, as stated above, the John Deere puller I think will do better work under all conditions than any I know of.

Again, will say in buying a puller, you want one that will pull ALL the beets.

Disc Before and After Plowing

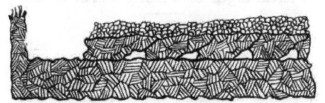

Figure 1. Showing effect of discing or harrowing heavy land after plowing. Note air spaces.

Figure 2. Showing how the disc closes the tops of the cracks and pores by breaking up the particles of earth.

Figure 3. Disced first, then plowed.

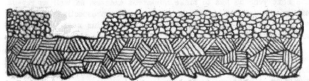

Figure 4. Perfected seed bed. Disced, plowed, and disced or harrowed.

Advantages of a Long Term Lease

Chas. W. Bowles
Landlord, Littleton, Colo.

In discussion of the question of the long-term lease or the short-term lease of farms, as between owner and tenant, the former has much the better of the argument when the interests of both are considered.

The customary time of the year for the new tenant to move onto the farm is March 1st, that being the time of the expiration of the lease of the former tenant. Instead of beginning his operations March 1st he should have been on that farm not later than August 15th of the previous year, in order to have begun fall plowing, or the preparation therefor, such as the irrigation of the land, if required, and if possible or necessary, the fertilization thereof.

The tenant with the yearly lease merely does the work necessary for the production of the crop of that year, refraining from the irrigation of hay land in the fall, which is very essential to the next year's crop, or the plowing and the preparation of the plowed land for fall or the next spring crops.

The one-year tenant does not know in August whether he will have the benefits of such labor the following year, and he does not feel like expending the money and energy for the benefit of his successor.

The man running the farm this year, knowing that he will have it for the next one—three or five years—often will make necessary repairs on the buildings, fences, ditches, private reservoirs, roads, etc., or the adjoining property, with little or no expense to the owner while the short term or one-year tenant will allow the fences, buildings and ditches to get in bad condition, weeds to grow, and is inclined to allow the farm to run down instead of trying to improve and build it up. When the owner shows such a farm to a probable tenant for the next year he has to make numerous excuses, as well as repairs of various kinds at his own expense, and the prospective tenant sees a run-down and uninviting looking farm for his next year's home and labors.

The tenant, after having irrigated a farm one year, has the advantage of a man who never has, as he will then have better knowledge of the conservation, use and handling of the water apportioned to or belonging to that farm.

Should the farm be of clay soil, much benefit is derived to both owner and tenant to have it plowed in the fall, as a good seed bed can then be easily obtained the following spring and better crops result.

454

The changing of tenants each year would cause this important work to be put off until spring, causing late seeding and oftimes poor crops.

(Editors' Note:—This is the first of a series of articles to be published in "Through the Leaves" on the "Advantages of a Long-Term Lease." These articles are to be written by both landlords and tenants. Anyone desiring to contribute an article on this subject send it to Longmont Factory.)

Some Questions and Answers

(From the "Utah Farmer)

Beet Top Ensilage

A number of farmers are going to silo their beet tops and crowns this year, if the interest shown in asking questions, is anything to go by.

The answers to these questions will no doubt help many other beet growers.

Ogden, Utah.

Utah Farmer:

Some people say we ought to put straw in alternate layers when siloing our beet tops. Has this practice proven satisfactory? Does the sizes, either in width or depth, make any difference for the beet tops to make good ensilage. Will freezing hurt the beet tops for feeding?

J. B.

Answer

Some excellent results have been had where straw is placed in alternate layers with beet tops in the pit silo. It is necessary, however, to do more packing, where straw is used, to keep the mass air tight. When the tops are gathered and put into the silo before the moisture has evaporated, it is entirely possible to make good ensilage when a six-inch layer of straw is alternated with twelve or fourteen inches of beet tops. The fresh tops have enough moisture in them to make the straw pack well. It is not safe to put in straw where the tops have shriveled and the moisture largely gone. Air is liable to enter the pit and cause the straw to mold.

The size of the pit does not matter seriously. We have seen excellent results where only a few tons were siloed. If the pit is made more than 14 feet wide, it is a little more difficult to cover with earth or pulp, in closing the pit. With the deep pit, the mass of tops and straw may be better packed than in the shallow pit. It is well to buid above the level of the surface especially if the water table is liable to rise and bother, where the pit is deep. Bank earth against the sides, as they are stacked and packed, to keep the mass from pushing outward, when ferment starts.

We have found good silage made from tops that had been frozen before the beets were dug.

* * *

Sugar City, Idaho.

Utah Farmer:

I am short of labor during the digging of my beets, what do you suggest to do so that I can get the best results? If beets are piled up and left for a week or two will they make good silage? Will putting them in the silo wet prevent them from making good ensilage?   —.S. J.

Answer

It often happens that piles of beets must be left on the field a week or two before they may be hauled. In that event, the beets should be covered with the tops to prevent the beets from freezing and also evaporating. The tops may be gathered and siloed after they have been laying in piles for two or three weeks. It is easier to pack the tops when put into the silo fresh. After the moisture has evaporated, the tops get spongy and do not readily pack.

* * *

Wellsville, Wash.

Utah Farmer:

I will appreciate a little information about siloing beet tops.

Do you recommend putting straw on the floor or bottom of the silo, also on the sides? Why do you suggest chaff as a covering for beet tops? What other kind of coverings can be used?   —J. R.

Answer

Straw may be used in the bottom of the beet top silo and also against the side walls. It is not necessary, however, to do so.

For a finishing cover over the tops, chaff is better than straw as chaff will pack better and thus exclude the air better. One of the best coverings, in finishing the silo is to spread a layer of wet beet pulp. It will do no harm to put a light spread of straw or chaff over the pulp. Pulp is wet and weighty and will readily seal the silo air tight. The heat from the fermenting silo also aids in keeping the pulp from freezing.

Earth to the depth of 6 or 8 inches is rather commonly used to cover the pit with. Any of these coverings will do. The value of the cover, primarily, is to seal up the silo and keep the air out. That is absolutely necessary.

* * *

West Jordan, Utah.

Utah Farmer:

I am going to silo some beet tops. Would like your advice on

456

building silo? Should the sides be slanting or straight down? If ends are left open for driving in and out, how do you suggest they be built up when silo is filled? —R. S.

Answer

The beet tops will pack better in a pit with perpendicular walls. With sandy soil, the banks will need to slope, however. The ends of the pit must be sloped to allow the team and wagon to drive through for unloading and for packing the mass. Let the ends and tops be built above the surface with some slope, in finishing.

Earth should be packed against the sides to keep the mass solidly packed. Crown the top so that rain or melting snow will drain off. Drain ditches should surround the pit to carry away the surface water. —The Utah Farmer.

SEALING THE SILO

By prompt and careful attention to the sealing of the silo a very appreciable saving can be made in a single silo. If every silo owner would take the time and give special attention to this matter immediately upon completion of the silo filling, the aggregate saving would be many thousands of tons of good silage.

Within two days after all the corn has been put into the silo, the top of the silage should be completely leveled, and this leveling should be accompanied by a very thorough treading, taking especial care to have the outside edge well packed. If at the time of this treading one to five barrels of water are put on the top of the silage, it will compact the silage, exclude the air, and facilitate settling. All three of these operations are desirable for good keeping. Immediately on top of the silage put ten or twelve bushels of grain chaff or short cut straw. Wet this thoroughly and tread it down hard. Upon the top of this wet chaff make a liberal sowing of oats. The moist chaff and the heating silage will sprout the oats. They will grow in great profusion. Their roots will form a dense mat over the top of the silo. The chaff will decompose slightly and will form an almost air-tight blanket on top of the silage. With the air thus excluded only a very slight decomposition of the silage will take place.

By thorough attention to these details from five hundred pounds to a ton and a half of silage can be saved in every silo.

—A. C. Anderson, Michigan College of Agriculture.

"More Milk-fed Americans" says the Rural New Yorker—and science agrees. —Hoard's Dairyman.

Double plow your alfalfa ground this fall

Surplus Tractor Power Prevents Stalling

W. E. Dodge

T HE more tractor experience a man has the more certain he is that he needs surplus power to take him over hard and unexpected places.

The man who owns a tractor that develops fifteen horse power at drive wheel rim finds himself in a disagreeable situation when, due to degree of grade or character of soil encountered, his machine stalls. He cannot very well back, if plowing, to get a new start. Therefore, the only alternative is to lift the plows and leave a blemish in the plowed field. What he wants, and what he is bound to have some day, is a tractor that will furnish, at least for a short time, a large surplus of power to take him up steep grades, through hard spots in the soil, or both at once.

A horse has the ability to produce surplus power in large amounts for short periods. For a moment, while tugging in the collar, he may produce five times or more of the amount of energy that is expected of him at steady work. Who has not seen the powerful truck horse straining at his load to get it up the grade to the unloading platform? While doing this the horse is exerting many times the strength required of him on the main portion of the journey over level streets The driver who gets stuck on the road or in the street fails for the entire trip, due to one bad place in the road; the same is true of the tractor that has not enough excess power, which may be called on when necessary, to carry over the steep grades or through tough hard spots in the soil.

It is the ambition of progressive tractor builders today to produce machines, that in a degree at least, compare with the versatility of the horse. When the farmer understands that he can purchase a tractor that for all the working period will develop, say, fifteen horse power except for ten minutes when he must have twenty-five horse power—that is the tractor he will buy and recommend to others when he finds he can get it.

There is a wide range of soil condition and soil content. It may take 2,240 pounds pull on the level to operate a fourteen-inch plow eight inches deep in gumbo soil, while the same plow may be pulled through sand at the same depth on level ground with a pull of 336 pounds. Assuming that the farmer ran from sand to gumbo in the same field, he could not hope to get through his gumbo with power sufficient only for the sand or silt loam. What he wants and will have some day is the latent power to call on when he strikes the streak of gumbo, so that he can go right through it without let or hindrance.

It has been fairly accurately estimated that it requires forty-two horse power to pull three fourteen-inch plows eight inches deep through dry gumbo soil, while to pull the same outfit the same depth through new land or virgin soil consumes ten horse power. In medium heavy clay the amount required is seventeen horse power, and sand about seven horse power. It is plain, therefore, that the range of power needed is wide indeed, and that the successful tractor must be built to meet this condition, modified to meet engineering restrictions. Virgin soil is, of course, not plowed eight inches deep, and it would not be wise to build a tractor capable of plowing dry gumbo eight inches deep, if the tractor was to be used generally on soil much easier to work.

The resistance which soil sets up against being turned over by the plow is determined roughly as follows: The cross-section of the strip plowed, say with a fourteen-inch plow set eight inches deep, is 8 x 14 or 112 square inches of surface of cross-section multiplied by the number of pounds pull required per square inch to move the plow. In the case of gumbo this is twenty pounds; hence, your problem is 8 x 44 x 20 or 2,240 pounds; and for a three-plow gang the problem is 8 x 14 x 20 x 3 or 6,720 pounds pull. In sand the problem is stated thus: 8 x 14 x 3 x 3 or 1,008. Thus we have 6,720 pounds pull necessary for passing through gumbo and 1,008 pounds pull for negotiating sand. The entire range between these figures may be encountered in the same field of operations, although this is not probable.

The man who lives in a rolling country will also find himself confronted with a real problem in getting his tractor to pull the load while going up the hill. A five per cent grade, which is equivalent to three degrees, will result in 300 pounds additional energy to move both tractor up hill, assuming that the tractor weighs 5,000 pounds and the plows weigh a 1,000 pounds. If, then, the user encounters a heavy streak of clay, or even gumbo, while going up hill, the need of reserve power is all the more obvious. A ten per cent grade, or 5¼ degrees grade means that the tractor must generate 600 additional pounds pull at the drive wheel rim in order to move forward; and on a thirty per cent grade, or 16¾ degrees, the additional load imposed is 1,800 pounds, or a difference of more than twelve horse power from that consumed on level ground, traveling at the average tractor speed.

Therefore, the only conclusion that can be reached in the application of tractor power to farm problems is that the tractor must be able, at least for short periods of time, to generate a much greater amount of power than that at which it is rated.

—From The Farm Journal.

PERIODIC

THROUGH the LEAVES

OCTOBER, 1919

THE GREAT WESTERN SUGAR COMPANY

THROUGH
THE
LEAVES

OCTOBER, 1919

Published Monthly By
THE GREAT WESTERN SUGAR CO.

DAILY CALL PRESS
LONGMONT, COLORADO
1919

TABLE OF CONTENTS

Siloed Beet Tops

A great deal has been said in these pages about siloed beet tops and each year finds an additional number of growers using this method of preserving their tops. When the method is once practiced it is nearly always continued. This year is no exception in that we have learned of many who are planning to lengthen their feeding season by siloing their tops to be fed next spring.

This valuable feed should be bringing greater returns than it has been in the past. Now is the time to act if you are going to this year. Think it over.

N. R. McCREERY,
Manager.

Notes

H. Mendelson

Receiving Beets

The beets in most districts are not as mature as in normal years. Nevertheless, the company decided to ask the farmers to start the harvest Monday, September 29th, to allow the maximum length of time for the beet harvest. It is also intended to start the factories earlier than usual, although we run a chance that a little unfavorable weather will leave us without beets enough to keep running, which is quite an expensive matter in factories with a large daily pay roll.

Our factories are mechanically in better shape than ever. It is hoped that this and the newly inaugurated eight-hour shift will enable us to keep up a high daily slicing capacity during the receiving season, unless the influenza strikes us again.

This is of interest to the farmers in so far as the more beets we slice during the receiving season, the less beets will have to be piled.

It must be emphasized again, as in years previous, that piling beets is not a punishment wantonly inflicted by the company upon the farmers, but is due to the fact that beets of necessity are harvested faster than they can be sliced.

Storing beets after they are received from the farmers is a costly and undesirable procedure, but cannot be prevented.

We have built a number of new dumps eevn in our older districts to facilitate the delivery. About $130,000 has been spent to put the old dumps in good shape. Nevertheless, delays may occur.

The company is fully aware of the importance of a speedy harvest to the farmer and will do everything in its power to prevent unnecessary delays. Nevertheless, a great deal will depend on the good will and co-operation of the farmers themselves. If, as is the custom at a number of stations, the majority of farmers insist on arriving at the dump at the same early morning hour, congestion must result, particularly when all farmers insist on going over the dump instead of unloading some loads into the pile. The so-called pro rating system, whereby every farmer is guaranteed his share in the available car space, is used at many stations. Even at these stations about 20 per cent of the beets are delivered during the first hour of the day.

If, in your opinion, there is reason for complaint against the way the dumps are handled, please complain to the fieldman, agricultural superintendent, or manager as soon as possible. Everyone of them will be glad to remedy anything that can be remedied.

We hear of many complaints only during February and March, when the new crop is being contracted for. This is, of course, rather late to do anything.

465

Tare

We never have received a complaint about tare being too low, although due to the nature of the sampling, there must be a number of cases where the tare is found too low.

Every beet raiser has the privilege of inspecting the tare-room and investigating in detail how the tare is determined, and we urge everyone in case of dissatisfaction to do this. The company cannot possibly have any other policy than to pay for every pound of beets delivered, no more and no less.

It is sometimes possible to decrease the percent tare by insisting on better work in the field. While our fieldmen will look after this as much as possible, it is really up to the farmers to do this. As shown in last month's issue by Mr. Maxson and others, probably quite a number of tons of beets are left in the ground every year by the negligence of the labor and by inefficient work of beet pullers.

* * *

Supervision of Inexperienced Help

We have to harvest this year about 205,000 acres as compared with 147,000 last year. This means there are at least 58,000 acres being handled by inexperienced help. The company has provided this help as well as could be done. We hope that farmers will see their way clear to give this inexperienced help sufficient supervision to get the crop out in the manner most advantageous to all parties concerned.

* * *

An Unfavorable Year

In nearly all districts the season has been unusually unfavorable, due to the dry and hot spring, preceded by a dry winter, with an unusual deficiency of snow in the mountains and the resulting lack of irrigation water.

The Sterling, Morgan and Brush districts and the Scottsbluff district suffered very little, if any, from lack of irrigation water.

It has been observed by many that in the beginning of the growing season even repeated applications of water did not produce the vigorous growth a normal rainy spell does.

In many fields a good growth took place only during the last part of August and the first part of September, and even now some fields look as green as they look normally in the middle of August. Many other fields appear yellow, as if they were ripe, although the sugar percentage shows they are far from ripe.

Leaf spot appears more prevalent than in normal years.

There will be many fields where the belated vigorous growth of tops will lead the farmers and fieldmen to overestimate the prospective yield. This is particularly true in a year following a season with exceptional high yields, like 1918. Last year there were a number of stations in Colorado averaging more than 15 tons per acre.

Some of them had the highest yield per acre during the last ten years. Naturally expectations at such stations are rather high. We hope that farmers in such districts will fully realize the extremely unfavorable conditions of the past summer.

* * *

Results of a Favorable Year

At the Colorado stations around 15 tons per acre or more last year, 693 farmers with 10 acres or more each were delivering beets. Their yields were as follows:

```
  1 with 23 tons per acre and less than 24
  8 with 22 tons per acre and less than 23
  3 with 21 tons per acre and less than 22
  9 with 20 tons per acre and less than 21
 55 with 19 tons per acre and less than 20
 67 with 18 tons per acre and less than 19
 79 with 17 tons per acre and less than 18
 91 with 16 tons per acre and less than 17
102 with 15 tons per acre and less than 16
 96 with 14 tons per acre and less than 15
 74 with 13 tons per acre and less than 14
 47 with 12 tons per acre and less than 13
 31 with 11 tons per acre and less than 12
 19 with 10 tons per acre and less than 11
 10 with  9 tons per acre and less than 10
  1 with  8 tons per acre and less than  9
```

Total....693

About one-seventh of all beet growing farmers in Colorado were delivering at these high yielding stations. Even where conditions were as favorable as they undoubtedly were, 100 farmers out of 700 got less than 13 tons per acre, while the best 100 in this group got more than 18.5 tons per acre. Is this difference in yields mainly due to the difference in the quality of land, or to the difference in the quality of farming? It certainly is not due to the weather.

* * *

Price of Feeder Cattle

The price paid for the bulk of feeder cattle at Denver during October was:

```
In 1916..........................................$5.90 to $ 7.15
In 1917...................................... 9.25 to  10.30
In 1918...................................... 9.50 to  11.00
```

At present good feeders sell from $10.00 to $11.00, with a strong demand and a probability that good quality will cost more money. Those starting in the feeding game should remember that good buying is a very essential factor in the success.

Figure 1

Winning Prizes at the County Fair

A. C. Maxson

T HE judge of any class of exhibits at an agricultural fair is always praised by winners on the one hand and condemned by those whose exhibits do not win on the other. Obviously a judge cannot be both right and wrong in his judgment at one and the same time.

In making awards the judge must be guided by the rules of the fair controlling the giving of prizes; he must see that each entry complies with the premium list as to breed or variety, age, size or number of individuals making up the exhibit or the total weight of the exhibit, etc.

The only thing which the judge really decides is the real quality of the exhibit. In cases where the premium list calls for certain breeds or varieties, he should be familiar with these so as to be able to determine whether or not an entry belongs in the class in which it is entered.

Too many exhibitors attempt to place all of the responsibility for their not winning upon the judge, not realizing or accepting their own share of the responsibility in selecting the exhibit and properly entering it.

After having acted as entry clerk for several years, the writer can truthfully say that not more than 50 per cent of the exhibitors know the qualifications of an exhibit as given in the premium list, and that fully 25 per cent do not know how they want to enter their exhibit or for what premium they are competing. In spite of all this they blame the judge because they do not win, when their losing is due to the fact that they did not do their part in selecting and entering their exhibit.

In order to win an exhibitor must study the premium list; he must be familiar with breeds of animals and poultry and varieties of field crops and vegetables; and last, but by no means least, he must know what constitutes a typical animal of the breed he is exhibiting and a typical specimen of the particular field crop or vegetable he is entering.

The real value of a fair to a community lies in the opportunity which it affords to study breeds of animals, poultry and varieties of field crops and vegetables. A study of the winning exhibits and a comparison of these with those that did not win is always profitable. Not being able to place the actual exhibits before the readers of "Through the Leaves," we are reproducing photographs of winning

469

Figure 2

Figure 3

Figure 4

Figure 5

Figure 6

471

Figure 7

Figure 8

Figure 9

Figure 10

473

Figure 11

Figure 12

474

Figure 13

Figure 14

475

exhibits and some of those that did not win, hoping that it may assist
our readers in selecting exhibits next year.

According to the premium list all beet growers could compete for
four prizes: First, for five best shaped sugar beets; second, five
largest sugar beets; third, five sugar beets with highest sugar content; and five best sugar beets, size, shape and sugar content considered. In this last twenty points were allowed for shape, forty for
size, and forty for sugar content, the prize being a loving cup.

Fig. 1 represents the winning exhibit and the cup. Comparing
the beets in Fig. 1 with those in Fig. 2, which were also entered for
the cup, is it any wonder that the judge did not give the prize to
the latter. Such ill-shaped beets would lose out entirely on shape.

Figure 15

On the other hand, coarse beets such as these seldom carry a very
high sugar content.

Fig. 3 represents the five best shaped beets in the fair. They
were not entered for the prize for shape, but are shown in order to
allow the reader to compare them with Fig. 4, which represents the
best exhibit entered for this prize. The judge gave Fig. 4 a third
prize, no first and second prizes being awarded in this class because
of the poor quality of the exhibits.

Fig. 5 was entered in the high sugar content class. Fig. 6 won
the prize. The coarse, ill-shaped beet stands very little chance of
winning in this class.

A premium was offered for two pie pumpkins. Fig. 7 won easily ever Fig. 8. Pie pumpkins are small and rather flattened at the stem and blossom ends. In Fig. 8 we see one good type pie pumpkin (left) and one with a regular field pumpkin type (right). The great difference in size of the two individuals detracts greatly from the quality of the exhibit also.

Fig. 9 represents two exhibits entered for the prize for best five pounds of table turnips. What housewife would not select the small, uniformily shaped turnips in the lower row in preference to the three coarse, irregularly shaped ones in the upper row. The judge awarded the prize to the lower ones. Do you blame him?

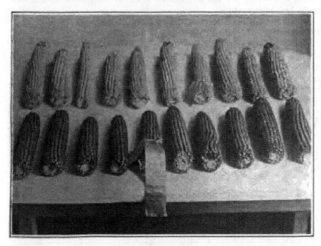

Figure 16

Several entries competed for the prize awarded the best five carrots for table use. The upper row in Fig. 10 won over the lower. Lack of uniformity in the size of the individuals in the lower row caused them to lose.

A prize was given for the best six cucumbers from 3½ to 5 inches long. The upper row in Fig. 11 won. The reason the lower row did not win is too obvious to need explanation.

Figs. 12 and 13 represent two entries of Early Ohio potatoes. Fig. 12 won easily over Fig. 13. Any housewife can tell why without asking the judge. Better let your wife pick out your potato exhibit.

Fig. 14 represents the only exhibit of Russett Burbank potatoes at the Boulder County Fair. The roughness of the individuals making up the exhibit caused the judge to place a third prize on it.

The prize for two best heads of cabbage, any variety, was given to the two heads at the right of Fig. 15. The judge placed a card on the two heads at the left, calling attention to the fact that in all probability two varieties were represented by this entry. This and the lack of uniformity in form of the two heads caused this entry to lose the prize.

Fig. 16 represents the ten best ears of Minnesota No. 13 yellow dent corn. The lower row won. Even though the upper row had been the only entry, it could not have won even a third prize, because of the greenness of the grain. The peaked tips and damage by worms at the tips would prevent this exhibit from winning even though the grain had been mature.

The musk melons represented by Fig. 17 won over those in Fig. 18 because they were more uniform in form and size.

In selecting exhibits for the fair next year, first study the premium list, make up your mind what prize you want to compete for and then be sure that your exhibit conforms with the requirements as given in the premium list as to number of individuals, weight, variety, etc. Lastly, select all individuals as near alike as possible in form, color and general appearance.

Figure 17

Figure 18

Five Year Average is Guide for Fall Wheat Plantings

ORE wheat should be sown this fall than was the average in prewar years, but not so much should be sown as was sown last year. This is the outstanding fall-farming recommendation of the United States Department of Agriculture, which is watching the changes of world supply and demand while European countries are getting back to normal in food production, and thus affecting the market for American products. The department's suggestions are based on the observations of specialists who were sent abroad to report on foreign conditions and probable needs, and on the most extensive reports it has been possible to obtain from other sources in this country and other countries.

Five-Year Average Safe.

As to winter wheat, the department suggests that 42,000,000 acres be sown this fall to this crop, and that 20,000,000 acres be sown in 1920 to spring wheat, making a probable aggregate production in 1920 of 830,000,000 bushels, of which 200,000,000 bushels would be available for export after home needs are met. This production would approximately equal the average yield of wheat in the United States for the five years 1915 to 1919, inclusive. The five-year average is thought to be a safe guide for American farmers.

The suggested acreage for fall-sown wheat is approximately 85 per cent of the area sown in the fall of 1918 and is about the same as was sown in the fall of 1917. The suggested area for spring wheat is approximately 88 per cent of the area sown in each of the last two years. The combined acreage of winter and spring wheat suggested for 1920 is about 86 per cent of the acreage sown for the 1919 crop, slightly more than the acreage sown for the bumper crop of 1915 and about 4 per cent less than the area sown for the 1918 crop.

The pressure of war demand for rye has ceased and it does not seem probable that exports next year will greatly exceed 25,000,000 bushels. Consumption of rye in this country was stimulated by the war to about 50,000,000 bushels. If this rate of consumption is maintained a crop of 75,000,000 to 80,000,000 bushels would be necessary to provide 25,000,000 bushels for export, indicating a total acreage of 5,500,000 to 6,000,000 acres, in comparison with 6,800,000 acres sown last fall.

"In any consideration of the probable foreign demand for American wheat and rye based on pre-war consumption experience," says the department, "a greatly increased transportation cost and the existing condition of international exchange cannot be disregarded. * * * It seems clear that if the foreign exchange situation continues as it is or becomes more abnormal it will constitute a powerful stimulus for some foreign countries to seek grain in countries other than the United States, or to produce it."

It is probable that several years will be required for European countries to get back to their normal cattle population. In order to supply their needs it will be necessary for them to import dairy products, and at the present time this country is supplying large quantities. Last year's exports were so large that about 2,000,000,000 pounds of milk were required to make the products, or 100 pounds from each cow in the United States. The department expects that exports will decrease and imports will increase with the exception of condensed milk. It recommends that dairymen give particular attention to economical feeding, through the buying of concentrates in large lots or the co-operative buying of feeds; attention to pastures that have been allowed to run down; attention to the feeding needs of their cows by the use of cow-testing associations; and the best use of labor and labor-saving devices.

Live Stock.

"What our foreign trade in meat and meat products will be in the future is impossible to predict with any degree of accuracy, owing to many factors which may affect it," the department says. "The exports will probably decrease as compared with the past year, as the European countries increase their production of live stock toward the pre-war production."

Other factors affecting American export trade are rates of exchange, which in many cases are decidedly unfavorable to those countries most in need of our meat and meat products; campaigns being waged in the European countries for increased production and decreased consumption; prices of meats and meat products in this country as compared with the prices in other countries having meats for export; available ocean tonnage for shipping from other countries.

However, pork exports at least probably will be greatly in excess next year of the pre-war annual exports. The European countries probably will require two years to get back to pre-war pork production, and their present needs are great.

Pre-war production of poultry in most European countries, the department says, will soon be resumed. Emphasis in this country is placed on more efficient methods of production. Whether increased production should be undertaken must necessarily depend on local conditions as to feed, labor, and other factors.—Weekly News Letter, U. S. Dept. of Agriculture.

Educated Youth has the Advantage

The value of staying at school is stated in dollars and cents in figures recently compiled by the bureau of education and distributed to boys and girls throughout the country by the children's bureau.

From a study of a large number of actual cases it has been found that at 25 years of age the boy who remained in school until he was 18 had received $2,000 more salary than the boy who left school at 14, and that the better educated youth was then receiving more than $900 a year more in pay.

"This is equivalent to an investment of $18,000 at 5 per cent," the statement said. "Can a boy increase his capital as fast in any other way?

"From this time on the salary of the better educated boy will rise still more rapidly, while the earnings of the boy who left school at 14 will increase but little."

While wages have increased with the war, the proportions shown in a table of weekly earnings still hold true, the statement said. The boy who left school at 14 at the time the investigation was made received an average of $4 a week, his wages increasing each year to $7 a week at 18. The boy who remained in school until he was 18 began work at $10 a week. At 20 the salaries were $9.50 a week for the boy who left school early and $15 for his better trained competitor. At 25 they were earning $12.75 and $31 respectively, and total wages up to that time had been $5,112.50 and $7,337.50, so that the boy who remained in school had earned nearly 50 per cent more in eight years than the other had in 12 years.

"Children should stay in school as long as possible because education means better jobs," the children's bureau urges. "Boys and girls who go to work at the end of grammar school rarely get good jobs. The work they find to do is usually unskilled; it offers little training or chance for advancement. When they are older they find they are still untrained for the skilled work that offers a future. Education means higher wages.

"Many boys and girls when they leave school find work that offers a high wage for a beginner. But these wages seldom grow because the work requires no training.

"A position with a future and steadily increasing wages requires school training.

"Does it pay to continue your studies? Education means a successful and useful life; it pays the individual. Education means efficient workers; it pays the nation. Show this to your parents and ask them what they think about it. Stay in school."—C. A. C. News Notes.

481

Improving Farm Livestock

P. H. McMaster

The United States Department of Agriculture has inaugurated a nation-wide campaign, beginning October 1, for the improvement in production of the farm livestock.

All previous endeavors to encourage improvement in the domestic animals has been largely ignored by the average farmer, for the reason that he has not considered such campaigns applicable to his own conditions. On the other hand, he has considered them as being applicable only to the pure-bred breeders on a large scale.

Regardless of the number of animals kept, the Department of Agriculture is desirous of interesting every farmer in this crusade.

It is readily realized that it is not possible for every farmer to produce pure-bred livestock. However, it is generally agreed that a great improvement can be made by using our present holdings as a foundation upon which to use pure-bred sires. Again, it is impossible for the small farmer to own privately an expensive sire.

The present stock, whether it be one or six of either hogs, horses, cattle or sheep, should be utilized in upbuilding your herd by crossing with a high production sire.

The production of sires is measured differently from the female species. In the case of horses a productive sire is one which is prepotent, or has the ability to transmit to his "get" his own characteristics. In the case of a dairy bull, his ability as a sire is measured by the milk production of his daughters, etc.

The plan to be followed in this improvement is a community-owned sire. Several farmers in each neighborhood buy a sire together and he is used by each on their respective animals. This plan has been followed for a number of years in the East, especially in Wisconsin where the dairy industry thrives.

The advantages of such a system of breeding are many, some of which are, no large investment is necessary for female foundation stock; the cost of sire is divided proportionately among several men according to the number of females to be served; it does not take long by judicious mating to improve your farm livestock. It only requires seven top crosses to produce a 99 per cent pure-bred.

The following illustrations show yearly averages of milk and fat by scrub cows and some of their daughters (by purebred bulls) used in a test at the Iowa State College to show the influence of prepotent sires on dairy production:

Fig. 1. Scrub, with an average production of 3,874.6 pounds of milk'
and 192.69 pounds of fat.

Fig. 2. Half-bred Holstein-Friesian (6,955.5 pounds of milk and
266.25 pounds of fat), a daughter of Fig. 1.

Fig. 3. Three-fourths-bred Holstein-Friesian (12,804.2 pounds of milk
and 4825.4 pounds of fat), a daughter of Fig. 2.

Fig. 4. Scrub, 4,338.5 pounds of milk and 183.49 pounds of fat.

Fig. 5. Half-bred Guernsey (4,213.1 pounds of milk and 179.72 pounds of fat), a daughter of Fig. 4.

Fig. 6. Three-fourths-bred Guernsey (6,582.6 pounds of milk and 333.11 pounds of fat), a daughter of Fig. 5

484

Fig. 7. Scrub. 3,313.2 pounds of milk and 178.47 pounds of fat.

Fig. 8. Half-bred Jersey (6,126.4 pounds of milk and 348.98 pounds of fat), a daughter of Fig 7.

Fig. 9. Three-fourths-bred Jersey, a daughter of Fig. 8.

485

The Long-Time Lease From the Tenant's Standpoint

A. F. Finley

 I HAVE farmed in Colorado for the past sixteen years, fourteen of which have been as a tenant farmer and I have always made it a point to secure at least a three-year lease if possible. There are several reasons why a tenant shoud do so.

First. A tenant moving to a new farm usually gets possession during the month of February, the worst month of the winter. The fences are in bad repair, the buildings likewise, and if he is not careful it will be the middle of March before he gets to do anything in the field. If he changes farms every year he will be from two to three weeks behind with his spring work. By a long-time lease this loss of time can be saved after the first year.

Second. It takes at least one year for any farmer, no matter how good an irrigator he may be, to become acquainted with the land so that he may irrigate it to the best advantage the second year, he will be able to correct his mistakes and the third year he just commences to reap the benefits of his experience and secure the maximum of production. All things being equal, the third year's profits will exceed the first two.

Third. Fertilization of the soil as we do it in Colorado, by the seeding of alfalfa and the use of manure, which are both expensive, holds no inducement for the short-term tenant. If he only has a lease for one year, he cannot afford to seed alfalfa or haul manure unless he is paid for his work by his landlord, as neither one will make much of an increase in the crop the first year. The second and third seasons are the ones that count.

Fourth. A one-year lease does not allow the tenant to engage in stock feeding, without which an irrigated farm, in my estimation, will only be partially successful. Farming and stock feeding are kindred industries, and one is incomplete without the other. And right here I wish to take a "poke" at the alfalfa meal industry; also the baling of alfalfa and shipping Colorado's most valuable product to foreign districts to be fed to stock and enrich soil hundreds of miles away. Farmers of Colorado, this is simply robbing the earth, you who are selling the alfalfa off of your own farms are standing in your own light; the sooner you begin feeding your own hay (no matter how much you are offered for it) and spreading the manure

486

thus obtained, you will be able to see your bank account grow. I have in the past sold some hay to alfalfa meal mills, but it was because I was forced to do so when leaving a farm and was unable to feed it out.

The reasons I have set forth in the above paragraph should be enough in itself to convince any land-owner as well as a tenant that a long-time lease is beneficial to both parties.

Fifth. Fall irrigation, which is conceded to be one of the greatest benefits to any irrigated farm, holds no inducement to the one-year tenant; he gets nothing out of it, consequently does not do it; or, if he does, the work is not properly done and the water is generally wasted, and a waste of water is a waste of money for the land-owner.

Sixth. Last, but not least, "Three moves are as good as a fire." The more often a farmer moves, the closer he gets to the bankruptcy court and the poor house. Now days this is wrong. No tenant can move from one farm to another, transfer his household goods and farm equipment, be the distance only across the road, for less than $100, and longer distances accordingly. If he is forced to make a long move, and it becomes necessary to have a public sale, his losses may easily run into the thousands.

Experience has taught me that moving is expensive.

Advantages of a Long Term Lease

G. W. Roop, Landlord

In any farming community the advantages of a lease for a term of years are many, but in our irrigated sections such a lease is still more important because of the fact that the best results cannot be obtained unless one knows many things necessary to the successful irrigation of a ranch, and such knowledge can be completely gained only by the experience in the irrigating operation during one season. To illustrate this fact I will tell our experience on our own ranch the first year we operated it. One part of the ranch was to be planted to beets and the direction to run our rows was a thing to be considered, so as to make the best use of our irrigation water. A ditch running along one side of the field was an indication that the water would flow parallel to it, so we decided to run our rows that way. In looking over the land we could not see that the grade was any different a distance from the ditch than it was near it; but when we began irrigating our beets we found that such was the case, and we sustained a material loss by having to use too much water on a part of the beets to get the other part irrigated.

In most of the irrigated sections the crops are more diversified than in non-irrigated sections, and the proper rotation of said crops is a very important consideration. The average ranch is not uniform in fertility and right crops should be selected for a particular field, that could be profitably grown there. I have seen cabbage set on land that would not produce a profitable crop of cabbage, but the same land would have produced a bumper crop of tomatoes. The tenant does not and cannot know all of these conditions the first year he is occupying a ranch.

Perhaps the greatest benefit derived by the long-term lease to both landlord and tenant is the interest that will be manifest by the tenant in the preparation of the land for future crops. A good tenant (if he has a lease for a term of years) will try to keep the land fertilized; he will keep the fields clean and will let no weeds remain to seed the fields the next year.

Be careful in the selection of your tenant. See that he is honest and efficient, and if you are sure that he has these qualities make his lease continuous.

The Long Term Lease

D. M. Jay, Landlord

It is with some reluctance that I write an article for publication for the simple reason that I feel there are many in this vicinity who have had much riper experience in farming and leasing under Colorado conditions than I have had, but having been a land-owner for a long time I feel that what is true in one state as regards farm tenancy will in the main hold true in another state.

I am a firm believer in long-time tenancy under right conditions. I feel that in making a lease, while the landlord has a right to expect a fair return for what he has invested, he should not try in the contract to work a hardship on the tenant, and after the lease is made there should be a hearty co-operation.

There are so many reasons against a one-year lease, while the reasons are legion for a longer tenancy, and, to my way of thinking, for this long-time tenant and the landlord to enter into a partnership whereby the gain will be mutual. I am a firm believer in a crop rotation and in double discing our stubble fields as soon as the small grain is taken off, thus retaining the moisture, and this year of all

488

years is forcing us to see that this point is very essential, making it possible to plow early because the moisture is there, and after this plowing is done I would, in the absence of a sub-surface packer, run a disc over the plowing with the discs set straight, to fill up any air pockets that might be left. All of this work would be seriously hampered by a one-year tenant, while a man staying on from year to year would enter into this plan heartily; and then, too, he would know how to take advantage of any little peculiarities of your soil, because of previous experience. If there was any one spot that needed a little more fertilizer, he would supply it. He would know the better how to plant the crops so that they might be irrigated to the best advantage, and how to run his temporary laterals. In fact, such a man of proven worth would be much more valuable to a landlord than the best one-year tenant could be.

The Scully estate in Illinois is the worst example of Absentee land ownership with its arbitrary management. Its small regard for maintenance of soil fertility, and last, but not least, its one-year leases makes this estate a blot on Illinois agriculture.

Dr. Thomas N. Carver says between the absentee landlord and the one-year tenant, the rural community is robbed of all that makes life worth living.

KANSAS HORSE DISEASE

The Kansas horse disease, which broke out in Colorado near Hartman and Bristol about the middle of July, has now spread east as far as Holly, west as far as Rocky Ford, and north to the vicinity of Ordway. Reports also indicate that some trouble is being experienced in the eastern part of the state in the vicinity of Fort Morgan and Wray. It seems probable also that a few cases have appeared in the San Luis Valley. In these latter situations, the disease has, however, not proven serious. The experience in the epidemic seems to indicate that treatment given early is of great value, so that it behooves the stockmen to detect the cases as early as possible. The first symptoms are dullness and an exhibition of a wabbly gait on being moved. In these districts stockmen should see their horses at least three times a day and note whether any of the animals are showing these symptoms. If so, a veterinarian should be called at once, as it is only in the early cases where treatment seems to be of much value. Do not under any circumstances attempt to drench the animals as in most instances the throat is paralyzed and the medicine goes into the lungs. This results in a mechanical pneumonia which always means death.—I. E. Newsom, Colorado Agricultural College, Fort Collins, Colorado.

Increased Production

P. H. McMaster

NCREASED production per unit rather than an increase in units for production is a problem confronting the American farmer today.

For many years farmers generally have endeavored to acquire additional acreage for cultivation and thus increase their profits by increased production. Many farmers were quite successful in this practice when land was cheaper and profits seemingly greater, but under the present high prices of farm land this practice has ceased to be as successful as formerly.

The modern farmer, and by this term I do not necessarily refer to some college-boy-farmer nor to a man who is in the game as a pasttime, I refer particularly to the man who applies business methods in the operation of his farm, and has learned that for him to profit from his business it is necessary for him to increase production per unit and discard such practices as have not proved successful.

First it has been necessary for him to keep accurate records of his farming operations that he may know at the end of the year which have proved profitable and which have not. In the case of the latter it has been necessary to decide what is the cause of this failure and if possible find a remedy for it. If it cannot be changed to a profitable basis it must be discarded for something else.

For instance one of these farmers bought more land at a very high price. After farming this additional acreage under practically the same conditions as prevailed in previous years he finds he has not made as much money as he did on the smaller acreage.

His records give him the cause for this and he is able to remedy the condition. He finds the reasons for his losses are as follows: Buying more land, increased the total investment, upon which interest must be paid. In addition to this he was forced to buy more equipment, such as horses and implements. This also increased the investment, interest and depreciation factors. Previously he had done most of his own work, but additional acres required him hiring more outside labor. This makes a large bill chargeable against the crop without considering the farmer's additional labor. A larger number of acres naturally required more time to farm, and, as a result, he was unable to devote as much time per acre to his crops as he had been in the habit of doing on the smaller farm. In other words, he was always behind with his work instead of being ahead of it ready to avail himself of every condition that lends itself to better farming.

This latter condition lowered his production per unit. He asked a friend for advice and they concluded that the land so recently acquired should be sold and returned to the smaller farm. The surplus equipment was sold and an intensive effort made to increase the production per acre with wonderful success.

The labor income, which is the net income after interest on investment, depreciation and all expenses are paid, was one hundred ninety-two dollars on the larger farm. Pretty small wage for year's work.

Two years later this farmer had the following story to tell. He had established a definite crop rotation as a consequence of which he was able to distribute his work more evenly throughout the calendar year. It was possible for him, therefore, to have seed beds in excellent condition; as there was plenty of time for him to do what was necessary to put seed bed in best condition, and he always was the first man in the community to plant his crops, thus insuring a lengthened growing season. He usually cultivated and irrigated his crops while other farmers all around were just finishing with the planting. As soon as the grain was harvested he stacked his grain and started fall plowing, getting ready for another early start the following spring. Thus by applying better methods he had increased his labor income from $192 to $1,537 the first year. How had this been accomplished? Simply by producing 50 bushels of grain where formerly 30, 18 tons of beets where formerly 10, 3½ tons alfalfa where formerly 2, and 10 tons corn ensilage where formerly 6.

His records showed him immediately what it had cost to keep his horses a year. Figuring the number of days each had worked, it was found that two had passed the point of maximum production and were raising the cost of production. Young stock, a team of grade mares, replaced these and production per unit was again increased in this department, and colts were raised to replace non-producers as they developed.

This farm maintained three dairy cows to produce dairy products for the household. After weighing milk for a month and having it tested for butterfat several times, it was found that three cows were not producing as much milk as one good cow would produce, and yet, they were consuming three times as much feed as the one. It is hardly necessary to say that they were quickly replaced by one cow.

A few hogs were also maintained for household use and a few for market. There were very few for market as the brood sow didn't produce very large litters. However, it wasn't long until this sow was replaced by one pure-bred Poland China. More pigs per year, quicker maturing and pigs requiring less feed per pound gain were produced.

This farmer was able within a period of two years to increase his labor income from $192 to $2,352. This was accomplished wholly by increased production per unit. The units consisted of acres of land, cows, pigs and horses.

The agreeable and profitable phase of increasing production per unit is that the same amount of work will produce a 50 bushel crop as a 30, or 18 tons as a 10, or it takes no more feed for a high-producing animal than it does a low producer.

Livestock Market Letter

Ben Kemper
President Denver Livestock Commission Co.

Receipts of cattle the past three weeks at all markets have been heavier than last year. Quality of cattle common, not as good as usual, largely due to the drouth in Wyoming and parts of Colorado, Idaho and Montana. Markets all sharply lower.

We quote today, Monday, the 22nd, good beef cattle a dollar lower. Good dehorned feeders 50 to 75c lower than three weeks ago; horned steers $1.00 lower; fat cows $1.00 lower; some medium kinds as much as $1.50 lower.

This break is due to several causes, mainly the campaign on the high cost of living and the agitation of the legislation against the packers at Washington, heavy receipts and shortage of cars. We don't look for much change in the situation soon, but with the price feeders are bringing now, compared with what they were bringing a year ago, largely $1.00 to $1.50 lower on steers and 50 to 75c lower on cows, I don't hesitate to say in my judgment these cattle will make money if put on a feed of pulp or beet tops, or a combination of pulp and insilage. No doubt, the uncertainty of the future and the high price of corn in the East is keeping a great many cattle out of the feed yards and will continue to do so, but we are bound to see better spots in this market in the next four months, especially cattle that are fat.

Choice feeders selling today from $9.50 to $10.25; good grades from $8.50 to $9.00; horned steers largely from $8.00 to $8.60; good, fleshy cows from $6.50 to $7.50; good thin cows, suitable for 90 to 100 days' feed, from $5.75 to $6.25. We should think it would be wise where feeders are situated so they could, to hold their cattle on pasture until they could place them in the feed yards and place their orders with their commission houses as I wouldn't be surprised to see the good kinds sell higher by the latter part of October.

Crop Rotations in America

Delmar C. Tingley

DURING the early history of mankind their source of food was directly from nature. They fished in streams and gathered the fruits from the trees that were growing wild. The various animals of the forests were hunted to supply the meat. Whenever the supply was exhausted in one locality the tribe migrated to a new place. As the population increased, and natures supply of food decreased, it became necessary for man to domesticate the wild animals in order to furnish a supply of meat, and to pull the plows so as to cultivate the soil. Finally when one plot of land ceased to produce any more crops, this land was left to nature and the tribe moved on and began to cultivate new land until that became exhausted of plant food. This continued until land became scarce. Then the idea developed that if the land was allowed to lay bare a year it would come back to its old producing power. This went on for many years. It was the basis of Roman agriculture for a long period, being systematically alternated with field crops every second or third year. The English farmers of the thirteenth to the seventeenth centuries also placed a great deal of dependence on the bare fallow in his scheme of cropping. Rev. John Lawrence says: "Fallowing kills weeds by turning their roots to the air. It lays the land in ridges, thereby better exposing it to receive the nitrous influences of frost, wind, sun and dew. These influences all tend to sweeten and mellow the land.

Modern agricultural science reveals the fact that, while the bare fallow acts as a temporary stimulus to soil productivity, it is a practice that serves to hasten the ultimate unproductivity of a soil area. The green manure fallow the annual pasture, and the legume meadows are modern means of resting land, and the bare fallow method is fast going out of practice except to kill noxious weeds and in our western section to conserve soil moisture.

Long before science discovered why the modern means of resting land was superior to the old method, the fallow was being used by farmers. They knew they redeemed this land; but it was a mystery how it did so; they did not know of the nitrofying bacteria that lived on the roots of the lupines, beans and vetches, the leguminous crops which were seeded at that time.

Alfalfa, however, was the old stand by of the ancient agriculture. It was brought into Europe from Asia long before Christ and gradually spread all over Europe. It became a standard farm crop in Roman agriculture about the time of Christ. A Roman author by the

493

name of Columella, writing about alfalfa in the early Christian era says. "But of all legumes, alfalfa is the best, because when once it is sown, it lasts ten years; because it can be mowed four times; and even six times in a year; because it improves soil; because all lean cattle grow fat by feeding upon it; because it is a remedy for sick beasts; and because two-thirds of an acre will feed three horses plentiful for a year.

It was not until 1888 that scientific investigations discovered the relationship that exists between legume crops and the nitrogen gathering bacteria. With this discovery it cleared up the mystery of why legume crops had such unusual power as soil renovators. From that time until now science has added much to our knowledge of soil bacteria. But with this there is still much to learn about legume crops and there place in a permanent scheme of agriculture. However, we know enough about these crops as soil renovators. The Romans realized this two thousand years ago, and when one sees the thousands of acres of cotton, corn and wheat in the United States still grown in reunions systems of continuous cropping, it makes one wonder if the Romans were not wiser farmers two thousand years ago than we are today.

Systematic crop rotation, however, as we now define it, means the alternation of the grain, grass and cultivated crops on a certain area of land. As the bare fallow begin to be strongly disapproved of by the leading agriculturists of England, a systematic system of crop rotation was developed. It became advocated so strongly that the leases of the English lands began to contain rigid cultivation clauses; which required tenants to manure land; allowed only two crops to be grown in succession and removed from the land; and stipulated that land sown to clover, if fed off, or with turnips fed on some part of the farm, were not to count as crops. A rotation which was developed at this time was (1) turnips; (2) barley; (3) clover; (1) wheat; (2) beans; (3) oats.

Crop rotation is an important feature in farm management. During the early history labor was cheap and little attention was paid to the economical management of man labor, horse labor, or machinery. But the twentieth century has changed labor conditions, and now the labor, horse power, and machinery are important factors in profitable farm management. The orderly arrangement of fields, and seasonal distribution of farm labor, provided by a well-planned crop rotation system, are basic features of good farm management, and the maximum efficiency of farm labor and machinery.

We may now consider the crops that may be used in a rotation. The staple field crops may be divided into four general groups for the purpose of studying and planning systems of crop rotation—namely, grains, grass, cultivated and cover crop.

494

Grain crops includes such crops as wheat, barley, rye, oats, flax and millet. These are all grown mostly for the value of their seed. Sometimes field peas, wrinkled peas, and soy beans, when sown for seed production, are classed as grain crops.

Grass crops includes those crops that are most commonly used for pasture and hay to be fed to farm animals, and include timothy, brome grass, red top, blue grass, red clover, alsike clover, bumson clover, white clover, and alfalfa. This type of crops require very little cultivation, and at the time they occpuy the land with their many roots they fill the surface soil with vegetable matter.

Cultivated crops includes those which are so planted as to permit of inter-tillage during the growing season. Examples of this class are: Indian corn, sorghum, kafer corn, potatoes, cotton, tobacco, and sugar beets. These crops are sometimes called cleaning crops, because they may be cultivated, and by this it is possible to free the weeds from the soil. It also helps to aerate the soil and to pulverize it.

Cover crops is used to designate certain crops that are sown for the purpose of covering the land, so as to prevent erosion and prevent soluble plant food from leaching. They are also to prevent the soil from drifting and are planted in orchards to keep the roots of the trees from freezing. The chief purpose is to protect the soil. The crops most commonly used are crimson clover, rape, buckwheat, winter rye, soy beans, cowpeas, and the vetches.

We have many defects resulting from continuous cropping on the same ground. When a grain crop is planted and continued to be planted on the same ground, the plants soon rob the soil of its nitrogen, and the soil soon becomes baked and ceases to produce a good crop. Weeds will also accumulate and finally run out the crop. The chief difficulty resulting from a continuously cultivated crop is that constant stirring of the soil decomposes humus at a rapid rate and will get good returns, but the soil will finally be limited in its supply. When a grass crop is grown continuously on the same ground in time the roots will become so matted that the plants will become sod-bound, and this will greatly diminish the yield of feed. In some cases of clover, after being planted on the same ground, causes a condition known as "clover sickness." Alfalfa will reach the same condition if continually grown on the same ground. It is also difficult to control plant disease and insect pests when continuous cropping is practiced. Many pests tend to live on one certain type of a crop, and in fact will not bother any other crops. So if we continue cropping we will finally get the soil so infested with special kind of insects and fungus pests that it will be impossible for one to produce a simular crop for years. A good example of this is shown by the potato crop.

The chief advantage in a crop rotation are: (1) It distributes your farm labor and makes the farm income more certain by having several crops in case one fails. (2) It provides a better means of controlling weeds and insects. (3) Helps to restore moisture in the soil. (4) Increases the nitrogen content of the soil. (5) By the different depths of root feeding it allows the soil to be aerated.

Crop rotation in itself is not the cure for all unproductive land or the absolute way to profit from high priced agricultural land. It is, however, the chief factor in a combination of good farming practice that will maintain the productivity of the soil and around which intensive systems of farming may be developed that will yield the maximum crop value per acre at the minimum of expense. Crop rotation is to general field agriculture what the foundation is to the house, the solid base on which we may successfully rear a permanent superstructure designed in a hundred different ways according to our individual requirements and desires.

WHEAT RUST ON FOXTAIL GRASS

A very heavy infection of what is undoubtedly black stem rust of wheat on wild foxtail grass has been found southwest of Loveland.

It is a well-known fact that foxtail, or as the botanist calls it— Hordeum jabatum, takes several of the cereal rusts spread by the barberry bush. Lending special interest to this instance is the fact that just west of the foxtail were found several hundred barberry bushes heavily infected with barberry rust. In addition, just east of the barberries and, likewise east of the foxtail, was a field of wheat which showed black stem rust. A clearer case against the barberry was never found.

The barberry spread the wheat rust to foxtail. From either the foxtail or the barberry it was spread to the wheat field.

Destroy the barberr y and bust the rust.—Walter J. Roth, Colorado Agricultural College, Fort Collins, Colorado.

GOOD START—GOOD ENDING

Clarence Sellers, a 14-year-old farm boy of Sedgwick County, Colo., started out right in a pig club. He bought a registered weanling Duroc-Jersey gilt, and his judgment in selecting a good pig was shown when his sow, "State-Line Duroc," developed into an animal of excellent show-ring type. Her marvelous growth and thick velvet covering of flesh spoke well for the feed and care given her. This sow entered the ring at the Denver National Western Stock Show as a senior pig weighing 350 pounds, smooth as velvet, with lots of length, height, bone, and vigor.

The initial cost of the pig was $18, and the cost of her feed was $30; the premiums won totaled $35, and the present value of the sow as a breeding animal is reckoned at $125.

Increasing Horse Power Efficiency

Wayne Dinsmore

EN on irrigated farms in Colorado are fortunately situated for the use of most efficient horses. They know by experience that draft horses can work in irregular fields around irrigation ditches, and across the occasional wet spots that are found in all irrigated lands, much more satisfactorily than any tractors, so that it becomes purely a question of using that type of horses that will enable one to do the most work in the least time. With this, they have the problem of using them in the most effective way.

The necessity of keeping the land on predetermined levels makes the two-way plow the most popular one in the irrigated sections. To operate this plow steadily and effectively, four good horses are needed.

The work done by the Illinois Experiment Station and the Percheron Society of America in the year 1918 brought out some striking facts which all horse users should know.

First of all, it was found that it is impossible to work three, four or more horses abreast without creating sidedraft, which makes the plow pull from 15 to 35 per cent heavier than it should. This can only be obviated by working the horses strung out in pairs. When they are worked with one team in front of the other, it is possible to do away with sidedraft, and the plow pulls one-quarter lighter than it does with four abreast.

From this it is at once evident that no man who has the good of his horses at heart will continue to work them three or four abreast. He will instead work four on his two-way plows, and by so doing will do away with the wasted power due to sidedraft. He will also have ample power to plow steadily all day at any depth desired, provided he has the massive draft horses which are most efficient in farm work.

The objection often raised against the tandem hitch by farmers is that the horses that are ten or twelve feet or further away from the plow cannot pull as effectively as if hitched closer. This was true in the common tandem hitches, but has been eliminated in the White Multiple hitches, which were developed during the progress of the experimental work referred to. A caboose pulls no heavier at the end of a 40-car train than it does when hitched directly behind the engine, because the angle of pull remains the same; and in the White Multiple Hitch the angle of pull remains the same on all teams, whether they are hitched directly to the plow or some distance out in front; and this keeps the angle of traces the same on all the horses—namely, at right angles to the shoulders, so that the lead team can, and does, exert its power just as effectively as the team hitched next the plow. This is the only hitch ever used on strung-out teams that does accomplish this end, and it makes it possible to work any number of teams strung out tandem fashion without any loss of power.

I have been much impressed in a recent trip to the irrigated section near Longmont with the extreme desirability of the four-horse multiple hitch for the two-way plow generally used there. The tying in and bucking back system, which is a part of the White Multiple Hitches, should be used, but lines should be left on the wheel team for use in backing the plow into corners which can only be reached in that way; but the lines of the rear team can be left untouched at all other times, for the tying in and bucking back system automatically controls the team when they are pulling, and more satisfactorily than the lines will.

On the larger fields, which are nearly half a mile long, I consider that the regular two-bottom gang plows will come into use with the six-horse multiple hitch as soon as Colorado farmers become aware of their labor-saving value. With a 6-horse team and gang plow of two 14-inch bottoms, it is easy to plow six acres per day per man; but these big outfits should be confined to the straightaway plowing, and the irregular pieces on the sides and ends should be finished with the four-horse outfit on the two-way plow.

The eight-horse outfit on a three-bottom plow is advisable only on large fields which are square or rectangular in shape with furrows at least half a mile long. With this outfit one man can plow eight and one-half acres per day or fifty acres per week. Very few irrigated farms are adapted to its use.

One great advantage in having ample power is found in the fact that you can attach a disc harrow or single section drag harrow behind these large outfits and harrow the ground as fast as it is plowed. This breaks the clods up thoroughly and leaves a finely pulverized top-soil which checks surface evaporation.

The four-horse multiple hitch will also be found extremely valuable on the wheat drill, beet digger, or in any other field work where a four-horse team is needed; and where it is desired to pull two big disc harrows, operating behind each other at once, two four-horse eveners can be attached to the four-horse multiple hitch and eight horse hitched instead of four, giving an abundance of power to do a big job well and quickly.

The use of the best kind of draft horses is as important as using them well. The ideal system of obtaining farm power is to raise it. Farmers using ten horses in harness should plan to have enough mares to raise four colts per year. This permits of selling off a pair of geldings when they are 3; and the 3-year-old mares can be put into the farm work, and a pair of the older mares sold in the fall when they are 7½ years old. This enables a man to sell his mares ere they are 8, before depreciation sets in, and if the right kind of a sire is used, the young mares coming into work each year will be better than their dams which are to be sold.

This plan permits a farmer to sell off two pairs each year for $800 to $1,200 for the horses sold, and keeps his work teams young, vigorous and constantly improving in size and quality. Furthermore, the mares with colts at side are available when the work is at its peak and can run in pasture with their foals when not needed, for they are producing revenue for the farmer even when not actually needed in harness.

Most of the farms in the irrigated sections of Colorado are small enough to be operated, so far as team work is concerned, by the farmer and his sons; and this is the ideal condition for the production of good draft horses, for the successful use of draft mares on the farm requires that kindly attention and interested watchfulness which an owner will give, but which hired hands are apt to lack. Men who handle their own teams with the help of their sons are the most successful draft horse raisers in the world, and practically all of the noted prize-winning Percherons of America were bred and reared on such farms.

Colorado also has alfalfa of the first quality, and oats that are unexcelled; and these, with good pasturage, furnish the ideal ration for producing high-class draft horses; but pasture for the growing colts from foals to 3-year-olds is absolutely necessary. No man can succeed without it.

The selection of foundation stock deserves an article in itself, but the fundamentals are the selection of sound, well-proportioned animals not more than 3 or 4 years old, that are true in type, size, conformation, quality and action to the breed they represent; for a man will do better with four good Percheron mares than with twenty common ones. Not how many, but how good, should be the motto of every beginner; and until you can buy good pure-breds, content yourself with the very best grade mares you can buy; for it is better to own a real good grade Percheron mare than an inferior pure-bred.

Good draft horses, used with modern equipment which eliminates waste power and doubles the amount a man can do per day, will, at the end of five years, increase any man's bank account enough to send his boy through college; and children like farms where good livestock, well handled, is strongly in evidence.

The more efficient use of horses leads to the use of more efficient, more profitable horses and upbuilds the farm and one's bank account.

New Stallion Law

Full Text of Act Which Went into Effect June 22, 1919

Section 1. It shall hereafter be unlawful for any person, persons, company or corporation to stand any stallion or jack for public service in the State of Colorado without first having obtained from the State Board of Stock Inspection Commissioners a license authorizing such public service. Such license shall be issued by the Secretary of the said State Board of Stock Inspection Commissioners, who shall charge and collect a fee of Five Dollars ($5.00) for such license, and shall authorize the public service of such stallion or jack for the calendar year of the year issued and shall state whether the stallion so licensed is a pure-bred registered stallion or a grade stallion, as the case may be. All such fees so collected shall be kept in a separate fund to be known as the Stallion Fund under the care of the State Treasurer to be disposed of as hereinafter provided.

Sec. 2. Before any license shall be issued for the public service of any stallion or jack, an application shall be made therefor upon suitable blanks to be provided by the State Board of Stock Inspection Commissioners, and in case of a pure-bred stallion or jack, the application shall be accompanied by the certificate of registration and in case of a grade stallion or jack the application shall be accompanied by an affidavit sworn to by either owner or breeder on suitable blank form provided by the State Board of Stock Inspection Commissioners, giving all necessary information regarding the breeding of said grade stallion or jack, including the registry number of either parent and name of breeder.

For the purpose of this act, a pure-bred stallion or jack is one that is registered in a book, recognized by the National Registry Board, and a grade stallion or jack is one whose either parent is registered in a book recognized by the National Registry Board.

A stallion or jack whose parents are pure-bred but of different breeds shall be licensed as a grade.

The license shall state in large type that the animal licensed is a Pure-bred Stallion or Jack, or that it is a Grade Stallion or Jack, as the case may be, and it shall give the breed, age, color marking and name or names of breeder and owner.

Sec. 3. There shall accompany each application for a license a certificate from some licensed veterinarian of the State of Colorado or Federal veterinarian, stating that the stallion or jack sought to be licensed is not afflicted with any of the diseases or unsoundnesses known as roaring, ringbone, chorea (St. Vitus Dance, crampiness, shivering, string halt), bone spavin, bog spavin, specific ophthalmia

500

(moon blindness), curb (when accompanied with curby conformation, or any form of venereal or other contagious diseases. Any stallion or jack that is a ridgling or that is deformed or afflicted with any of the diseases or unsoundness above mentioned shall not be imported into Colorado to be used for breeding purposes, or if in the state, shall not be licensed for service in the state, and it shall be unlawful for any person, persons, company or corporation to stand any such stallion or jack for public service in this state; provided that in securing the certificate from any licensed veterinarian on the examination of any such stallion or jack sought to be licensed, it shall be unlawful for any licensed veterinarian to charge more than Three Dollars ($3.00) and actual expenses as fee for such service.

Sec. 4. Should the attention of the State Veterinary Surgeon be called to the fact that any stallion or jack being stood for public service under a license as provided herein, is afflicted with any of the diseases named in Section 3 of this Act, it shall be the duty of the State Veterinary Surgeon to make an examination of such animal, and if the animal is found to be so afflicted, it shall be the duty of the State Veterinary Surgeon to report the same to the State Board of Stock Inspection Commissioners, who shall revoke the license of said animal and cancel the same, and thereafter such stallion or jack so afflicted shall not be allowed to stand for public service in this State.

Sec. 5. The State Board of Stock Inspection Commissioners is authorized and instructed to use the money in the "Stallion Fund" each year for the purpose of enforcing this act and of offering special premiums at any fairs or stock shows held in this State to be competed for by horses, mules or colts owned in Colorado and exhibited at such fairs or stock shows, and said Board shall make such rules and regulations in regard to such premiums and the competition therefor, as will in their judgment best encourage the breeding of improved horses and mules in Colorado and the premiums offered shall be competed for under such conditions as will best produce that result. The State Board of Stock Inspection Commissioners shall at the close of each year, providing funds will permit, issue and distribute free of charge, a report of all stallions or jacks licensed during the year, giving their ownership, breed, age, location; the cost of printing and distribution shall be paid for from the "Stallion Fund." Any money remaining thereafter in said "Stallion Fund" may be used as the said State Board of Stock Inspection Commissioners see fit in furthering the horse industry of the State.

Sec. 6. Any person who shall offer the breeding services of any stallion or jack to the public in this State, or make a charge for such service, shall have a license as provided herein and shall keep such license posted in a prominent place on or near the stall where the said stallion or jack is kept, or if the stallion or jack is upon the road being vended for service, then the person in charge of said stallion or jack shall carry or exhibit such license to the owner in charge or care of any mare to be served by said stallion or jack, and such license must be shown for the inspection of any person desiring the service of the stallion or jack, or who may for any reason desire to see such license.

Sec. 7. The owner or owners of any licensed stallion or jack shall have a lien upon the get of any such stallion or jack until the service fee has been paid, and it shall be unlawful to sell or dispose of or remove from the county any colt upon which the owner of a licensed stallion or jack has a lien as provided for in this section without the written consent of such owner, nor shall any mare served by a licensed stallion or jack be sold or removed from the county prior to payment of service fee without the written consent of the owner of the stallion or jack which served said mare.

Sec. 8. Any person, persons, company or corporation, who shall violate any of the provisions of this act, or who shall change the color markings of any stallion or jack, or give false verbal or written statements regarding the breeding of any stallion' or jack with the intent to deceive, shall be deemed guilty of a misdemeanor and upon conviction thereof in any court of competent jurisdiction shall be fined a sum not to exceed one hundred dollars ($100.00) or may be imprisoned in the county jail not to exceed thirty days, or both such fine and imprisonment, at the discretion of the court. Any fines so collected shall be turned into the "Stallion Fund" under the care of the Treasurer of the State Board of Stock Inspection Commissioners.

Approved March 24, 1919.

The disc should precede the plow in most cases, especially in the fall breaking, to pulverize the hard surface so it will unite readily with the subsurface soil when the plow turns the furrow upside down. It also mixes the trash through the soil instead of turning it down as a loose mat to check the rise of water during the growing season. The disc is one of the most rapid and effective pulverizers we have for following the plow, either for breaking up clods or cutting sod. In some places it is used instead of the plow for preparing ground for grain or alfalfa, but this is a poor sort of farming unless the ground has been well broken for cultivated crops this year and is in fine condition except on top. When well sharpened it cuts up corn stalks and does away with the necessity for a stock cutter. It will maintain a dust mulch or renovate a pasture or meadow. It even makes a fairly good subsurface packer when well weighted and run with discs set practically straight. It can be used more days in the year than any other piece of heavy equipment unless it is a wagon or tractor.

What is Breeding

The following definitions have been adopted by the United States Department of Agriculture for use in the "Better Sires-Better Stock" campaign which it will conduct in co-operation with the various states, beginning October 1:

Pure-bred: A pure-bred animal is of pure breeding representing a definite, recognized breed and both of whose parents were pure-bred animals of the same breed. To be considered pure-bred, live stock must be either registered, eligible to registration, or (in the absence of public registry for that class) have such lineage that its pure breeding can be definitely proved. To be of good type and quality, the animal must be healthy, vigorous and a creditable specimen of its breed.

Thoroughbred: The term "thoroughbred" applies accurately only to the breed of running horses eligible to registration in the General Stud Book of England, the American Stud Book, or affiliated Stud Books for thoroughbred horses in other countries.

Standardbred: Applied to horses, this term refers to a distinct breed of American light horses, which includes both trotters and pacers which are eligible to registration in the American Trotting Register. Applied to poultry, the term includes all birds bred to conform to the standards of form, color, markings, weight, etc., for the various breeds under the standard of perfection of the American Poultry Association.

Scrub: A scrub is an animal of mixed or unknown breeding without definite type or markings. Such terms as native, mongrel, razor-back, dunghill, piney woods, cayuse, broncho and mustang are somewhat synonomous with "scrub," although many of the animals described by these terms have a certain fixity of type, even though they present no evidence of systematic improved breeding.

Crossbred: This term applies to the progeny of pure-bred parents of different breeds, but of the same species.

Grade: A grade is the offspring resulting from mating animals not pure-bred, but having close pure-bred ancestors. The offsprings of a pure-bred and grade is also a grade, but through progressive improvement becomes a high grade.—From Hoard's Dairyman.

The Cost of Filling a Silo

H. J. Metcalf
Iowa State College

Filling a silo is work which no one looks forward to with any great degree of longing and which calls for a goodly outlay of cash. However, the returns are such that the expenditure is more than justified.

One should not go at the job shorthanded. There should be at least one man with a binder in the field with a day's start so that there will always be bundles ready to haul in. At best four pitchers are needed. Little if any money is saved by buying an underslung wagon so that the driver may put on his load. While there is a slight advantage in favor of the underslung wagon while loading, there is a disadvantage when every bundle must be lifted up to the cutter. The ordinary tight bottom rack is preferable for all kinds of bundles and hay hauling on the farm. Men work on them better, there being no danger of falling through about the time they have a good load on a fork or on their arms.

In 1918 Prof. J. M. Evvard of the Iowa State College kept a close account of the labor cost of filling several silos on the college farm at Ames. These figures lead us to the belief that silo filling will cost $130 a day this fall, divided as follows: Eight teams hauling from field to silo at $6.50 per day, $52; four men pitching in field at $4 per day, $16; one man, team and binder in field, $15; one feeder at cutter, $4.50; two men in silo tramping at $4 and $8; one engine and engineer, $15; fuel, $6.50; incidentals—oil, grease, etc., $2; sharpening knives, $1; cutter—cost, interest, depreciation, risk, etc., $10.

If everything goes nicely, $13 an hour should put the corn in the silo, and with everybody on the job 80 tons should represent a good day's work. The labor cost per ton then is $1.63. It might be advisable to add a manager to the crew to make sure that everything is kept going. If enough time is lost so that only 70 tons are handled, the labor cost per ton runs up to $1.86; or, if cut to 50 tons, it soars to $2.60 per ton. —From the Breeders' Gazette.

Meteorological Report, Longmont

FOR AUGUST, 1918 AND 1919

TEMPERATURES:	1919	1918
Mean Maximum - - -	89.48°	85.45°
Mean Minimum - - -	51.29°	52.23°
Monthly Mean - - -	70.38°	68.84°
Departure from Normal - -	+2.18°	+0.70°
Maximum - - -	99.00° on 14th, 22nd	96.00° on 2nd
Minimum - - - -	45.00° on 12th, 30th	42.00° on 22nd
PRECIPITATION IN INCHES:		
To Date - - - -	6.62	15.12
For Month - - - -	1.62	1.89
Greatest in 24 hours - -	1.35 on 2nd	0.80 on 6th
Departure from normal for mo.	+0.45	+0.74
Departure from normal since Jan. 1 −4.93		
NUMBER OF DAYS:		
Clear - - - - -	15	7
Partly Cloudy - - - -	16	24
Cloudy - - - - -	0	0

STATE'S FARM ACREAGE

Seven Per Cent of Colorado's Total Area Now Under Cultivation, Bureau Reports

Practically complete reports from county assessors to the State Immigration Bureau show that approximately seven per cent of the state's acreage is in cultivation this year, not including the area devoted to wild hay. Revised figures show that Phillips County is the state's leading agricultural county in proportion to its size, having a larger percentage of its area in cultivation than any other county. Preliminary reports gave this honor to Logan County, which has under cultivation this year 405,736 acres, or 34.79 per cent of its total acreage. Phillips, which is a much smaller county, has 165,212 acres under cultivation, or 37.52 per cent of its area. Wheat and corn are the leading crops in both these counties and the yields of the former crop will be above the average. Eight counties in the northeast corner of the state, Weld, Adams, Morgan, Logan, Sedgwick, Phillips, Washington and Yuma, have 2,205,000 acres of land under cultivation this year, or approximately 50 per cent of the entire cultivated acreage. —Commercial.

Fall Plowing

The past season has caused nearly every farmer who has irrigated land to mention the fact that he neglected or missed an opportunity when he did not plow his land last fall. This harvest will show a loss in our grain crops of about eight bushels of grain per acre and about two tons of beets. The grain loss is worth $16 per acre and the beet loss about $20.00 per acre. In some instances the loss to sugar beet growers has exceeded fifty dollars per acre and in a few instances a total failure. A number of beet fields have failed in the germination from first planting, causing the farmer to irrigate his field and resurface the seed bed, which is a heavy expense at prevailing prices and usually a less tonnage is secured. To overcome the possibility of not being able to do the necessary fall plowing and suffer the consequent crop loss the following season, it would be money well spent to hire the plowing done, by securing someone with a tractor who would be pleased to secure custom work by plowing ten, twenty, fifty or more acres for different men and have the land aerated by the elements during the freezing and thawing of winter. This season has demonstrated that most of our spring plowing should have been done last fall, and our crops would have been increased sufficiently, that the cost of fall plowing hired done, would have remunerated us enough in additional money to have had the plowing done gratis and a bank credit besides. In the neighborhood of 100,000 acres plowed last spring in Utah, we will sustain a crop loss of nearly $2,000,000, which we might have secured had we hired additional teams and tractor help. There are a number of men who are willing to do custom tractor plowing, and this method should be encouraged.

Another feature coupled with fall plowing is to irrigate the land before plowing in the fall, provided there is an insufficient amount of moisture to guarantee deep and mellow plowing.

Where there is sufficient water in the fall to irrigate dry land grain fields that are fairly level and no water available for summer use, fall irrigation has increased the wheat yield on otherwise dry farm land 25 per cent.—John Holmgren, Garland, Utah.

Boulder County Fair

The Fair Beautiful

As Seen by the Camera

LONGMONT, COLORADO, SEPT. 2-5, 1919

"FARM PRODUCTS" EXHIBIT, BOULDER COUNTY FAIR

256

"FARM PRODUCTS" EXHIBIT, BOULDER COUNTY FAIR

"FARM PRODUCTS" EXHIBIT, BOULDER COUNTY FAIR

EXHIBIT FURNISHED BY COLORADO COLLEGE OF AGRICULTURE TO FURTHER ERADICATION OF COMMON
BARBERRY

DUROCS GAINING PROMINENCE

Critic Model; 1st in class under 6 months; 2nd to Grand Champion, any age or breed, owned and exhibited by Burt Hart.

Miss Gano King 2nd, Champion Sow, any age or breed, owned and exhibited by Burt Hart, Longmont.

Bergman's Iowa King (10845); owned and exhibited by A. F. and A. H. Bergman, Longmont. First premium boar over 6 months and under 12 months.

Big Bone Wonder, Jr. (16523) and Sunset Wonder (16239); owned and exhibited by A. F. Bergman, Longmont. The former won first premium in class for boars 6 month and under.

1st Premium Aged sow Owned and Exhibited by J. A. Davidson, Longmont, Colo.

Longmont King (61116) owned and exhibited by J. A. Davidson, Longmont, Colo. This boar was Reserve Champion, Western Livestock Show, Denver.

Victory, (235801); Grand Champion Poland China Sow, Owned and Exhibited by A. D. McGilvray.

Sampson Over, (96987); Grand Champion Boar, Owned by A. D. McGilvray.

DUROCS GAINING PROMINENCE

Critics Model; 1st in class under 6 months; 2nd to Grand Champion, any age or breed, owned and exhibited by Burt Hart.

Miss Gono King 3rd; Champion Sow, any age or breed, owned and exhibited by Burt Hart. Longmont.

Longmont Gono King (16436); owned and exhibited by A. F. and N. H. Bergman, Longmont. First premium boar over 6 months and under 12 months.

Big Bone Wonder Jr. (16193), and Sunset Wonder, (16179), owned and exhibited by A. F. Bergman, Longmont. The former won first premium in class for boars 6 month and under.

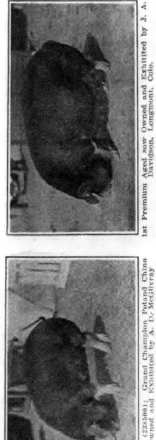

Victory, (238500); Grand Champion Poland China Sow, Owned and Exhibited by A. D. McGilvray

1st Premium Aged sow Owned and Exhibited by J. A. Davidson, Longmont, Colo.

Longmont King (91116), owned and exhibited by J. A. Davidson, Longmont, Colo. This boar was Reserve Champion. Western Livestock Show. Denver. January 1919.

Sampson Over. (96987); Grand Champion Boar, Owned by A. D. McGilvray

Scottish Lady 4th (696456); Grand Champion Female Shorthorn, Any Age. Owned and Exhibited by H. P. Harmon, Boulder, Colo.

Beauty's Clipper (695065). Champion Shorthorn Bull, over 12 months, special premium offered by Short-horn Breeders Assn., owned and exhibited by J. M. Winslow.

Divide Wildflower (646619); First in class for Short-horn Heifer, 2 years and under 3. Owned and exhib-ited by Glen E. Martin.

From Right to Left—Parkdale Victoria 6th, (167695); First prize Shorthorn Cow over 3 years, owned and exhibited by Everett Harmon, Boulder, Colo. Lustile (152300); 2nd prize Cow over 3 years, owned and exhibited by Glen E. Martin, Boulder, Colo.

Four Horse Hitch, Owned and Exhibited by Fred Muhme, 1st Premium

From Right to Left—Pure bred Percheron Mares, owned and exhibited by L. F. Bein, Berthoud, first premium. Grade Mares owned and exhibited

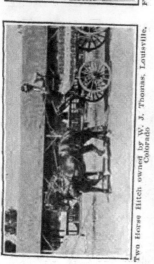

Two Horse Hitch owned by W. J. Thomas, Louisville, Colorado

Two Horse Hitch, Owned and Exhibited by L. F. Bein, Berthoud, Colo, 1st Premium

Pure bred Shire Colt, Grand Champion, bred, owned and exhibited by R. C. Miller, Lafayette, Colo. This colt is sired by Moultons Chief (116881); Dam, Starlight Daisy (14348). This colt won special premium offered by Longmont Farmers Milling and Elevator Co. Picture of cup is here shown.

Champion Grade Mare, any age. This Filly is a "Short" Two Year Old.
She was Bred, Owned and Exhibited by Clinton Eversole.

Joffre, 2nd in Class for 2 Years and Under 3, Owned and Exhibited by
Wm. Hansen, Longmont, Colo.

Quinton (134301), First Premium for Percheron Stallion, 3 Years and
Under 4, Owned and Exhibited by L. F. Bein, Berthoud.

Moulton's Chief (16881); Shire Stallion, Owned and Exhibited by R. C.
Miller, Lafayette.   First Premium Shire Stallion in Class,
4 years and Over.

FIRST PRIZE SHROPSHIRE EWE ON LEFT. SECOND PRIZE SHROPSHIRE LAMB ON RIGHT. EWE IS DAM OF LAMB. BRED, OWNED AND EXHIBITED BY MRS. DAN BURCH, NIWOT, COLO.

DAIRY CATTLE GAINING IN POPULARITY

GUERNSEY HERD, OWNED AND EXHIBITED BY BENDER & SONS, JOHNSTOWN, COLO.

ORMSBY SKYLARK JOHANNA (175068): GRAND CHAMPION HOLSTEIN BULL, OWNED AND EXHIBITED BY J. J. KALBERER, BROOMFIELD, COLO.

Victor De Furnes (9390); First Premium Belgian Stallion, Owned and
Exhibited by Dick Beasley

First Premium Duroc Sow and Litter, Owned and Exhibited by Bart
Hart, Longmont.

THROUGH the LEAVES

NOVEMBER, 1919

THE GREAT WESTERN SUGAR COMPANY

THROUGH
THE
LEAVES

NOVEMBER, 1919

Published Monthly by
The Great Western Sugar Company

Longmont Call Press
1919

TABLE OF CONTENTS

Beet Growers and Families from Sable District Visit Longmont Factory

Editor's Page

N. R. McCreery

This year's beet harvest is practically completed, and in view of the general labor unrest throughout the whole world, the Northern Colorado farmers fared very well indeed. Few of you realize the effort that was made on the part of the company to supply this labor, and not knowing intimately, few perhaps appreciate it to the extent they should. We realize that in many instances the labor was not all that could be desired. We know, however, that if we could have been advised thirty or forty days earlier of your requirements, we could have done much better in the selecting of this help. There is every indication at this time that next year will see a large acreage of bsets contracted. We ask that you remember this year's experiences and let us know very soon after the first of the year what your requirements will be.

Some are also inclined to think that the fault is always with the labor. There are many instances where this is true, but there are also many instances of unfair dealing on the part of the grower. Too many growers expect the labor to grow their crop for them. They do not do their part to make the thinning easy by proper cultivation before thinning, and then they allow a growth of weeds to come in between the rows, and because the field looks dirty they blame it on the labor. They also come to the harvest with a desire to get their crop out of the ground as quickly as possible. This is well and good, but some consideration must be given to the labor who has stayed all summer to do the fall work. You may have equipment enough to finish your harvest in two or three weeks, but unless you are willing to pay a much larger contract price you cannot expect a laborer to have enough people to do this work in so short a time. In other words, a reasonable time must be allowed before you demand that more help be put in the field. True, there are instances where there is not sufficient help and these must be helped out, but let us figure the matter fairly and then act in a perfectly fair manner.

Then there is the matter of housing. During the past year or two a great deal has been done in the way of building better houses, but there are still a large number of shacks that are not fit habitations for human beings. More and more the labor is asking the question: "What kind of a house is there to live in?" A suggestion has been made that may be carried out, that a photograph of the beet house be attached to each labor contract. We know of some growers whose contracts will be hard to fill if a picture of their laborer's house is attached. Landlords are receiving a sufficient return in most instances to justify a good beet worker's house, and the tenant should insist that this be furnished as a part of the farm equipment.

The facts are, that if we are to continue to grow beets, labor for the hand work must be secured aside from the American help. And if these people are to come back year after year they must receive fair treatment at the hands of the people they work for. We are not endeavoring to excuse or defend poor work; we know there are many instances of poor labor and poor quality work. We know also of instances of injustice on the part of the grower, and in view of general conditions that make it difficult and expensive to secure this help, we want to make a plea for tolerance and fairness on the part of the growers.

* * *

We are publishing in this issue two articles on fall-plowing, one somewhat contradictory to the other.

Any farmer fully realizes that it is impossible to make a general rule as regards farm practices. Fall-plowing is no exception. There are some kinds of soil which will produce better when plowed in the spring, rather than the fall.

The larger percentage of the beet growing land in Northern Colorado can be plowed in the fall and better results obtained therefrom. The farmers who own land that cannot be properly plowed in the fall know their conditions better than any other person, and they usually make proper allowances in any general plan that may be advocated.

Fall-plowing itself is not the cause of many failures. It is caused by not properly finishing the job. If the land is left in the rough, and we have a dry winter, the soil will dry out very badly. For this reason we advocate packing the soil with a disc set straight.

If land is plowed when dry and hard, whether it be in the spring or fall, it is next to impossible to make a good seed bed. Judgment must be used in all cases as to whether the land is in proper condition for plowing.

Fall plowing is much better for heavy soils; for sandy soils it is very seldom practicable to fall-plow because of danger of blowing. The question as to fall or spring plowing concerns mostly the silt and loam soils.

From Factory and Field

Greeley

The farmers of Northern Colorado contracted the largest acreage of beets that was ever contracted in Northern Colorado. You farmers prepared your seed beds as well or better than ever before, and in our early planting season our prospects were very good for a heavy tonnage. Early planted beets made good stands, as a rule. On the late planted beets, with no moisture, there were a good many fields and parts of fields that were lost; also a good many fields, with poor stands, so much so that a portion of them were abandoned. The drouth continued from the 10th of April until the 30th of August with only two inches of rain, also little water in the canals, and very little snow left in the mountains to supply our needs. Our crop looked very discouraging and very backward, and our prospects were poor for a good crop. In addition to the backward spring and the prolonged drouth, on about June 1st the web worms made their appearance. For a time it looked like the worms were going to get the best of us, and they did with a few farmers where they did not spray in time.

The farmers were aware of the danger and went to work with a will, and I want to say to you farmers, you saved your crop. I want to add to this that those of you who had water and applied the water to your beets directly after the foliage was eaten off the beets are making a good fair tonnage. The sugar is low, and naturally it would be low under these conditions. I just mention the past conditions to remind you how near a beet crop can be destroyed and revived and still make a good crop. I am almost sure from what beets that have been harvested that our beet crop this year will be as heavy if not heavier than in 1918.

Now our harvest is on and up to the present time we have harvested more than three times as many beets as we have sliced at our factory. We are receiving at the rate of between 5,000 and 6,000 tons per day, and it is the desire of this company that you continue to deliver your beets as fast as you see fit. It is also the desire of this company that you deliver your beets in a good marketable condition, free from dirt, and topped below the first or bottom leaf.

Sometimes I wish that there was a delay in harvesting beets, and especially a fall like this where there is plenty of moisture, and the land is in such beautiful condition for fall plowing. A great many farmers would be ahead, if they could stop harvesting for ten or twenty days, and fall plow for next year's beet crop, providing they

531

want to change the land. It is an undisputed fact that their land is in good condition for moisture, and fall plowed the land will produce from two to five tons of beets per acre more than with spring plowing, and you will raise your crop cheaper and with less work by reason of an early growth, which checks the growth of weeds and your beets will be thinned before hot weather, and the growth will not be retarded by reason of hot weather. Some might say that this is a little early to think of next year's crop, but it is not. Now is the time to prepare for next year's crop, and every farmer that will prepare now for next year's crop will make a thorough success of farming in Northern Colorado, for we have such beautiful land and with such thorough irrigation systems, with good climatic conditions there is no question about your success, providing you do your part in the right time and in the right way.

Let us continue to improve our soils so as to produce more tonnage, so it can be rightfully said that Colorado has the most fertile soils of any state in the Union, and prove it by raising more tonnage. Your land will continue to furnish more employment and will go higher and higher in value from year to year, and God speed the time when every twenty acres of irrigated land in Colorado will support a family, and every family will own their own home, then is the time when we will be a prosperous and happy people.

H. TIMOTHY, Agricultural Superintendent.

Billings

The beet harvest is progressing satisfactorily and two weeks more of good weather will see us through, while the acreage (owing to the severe drought) is small, the yield will be somewhat better than expected.

The only good yields of wheat in this territory this year we have heard of were grown on last year's beet ground in the vicinity of Ballantine on the Huntley project, where yields from 40 to better than 60 bushels were reported. In all other sections from 10 to 20 bushels were the rule, and a good deal of that wild oats, due, no doubt, to the irrational practice of following with wheat after wheat without any rotation, until the land refuses to grow any kind of a paying crop. How much better it is to rotate and get twice the yield and better quality.

Very little fall plowing has been done to date—a condition which is to be regretted, because any plowing done this fall means not only better crops next year, but much work saved in the spring when every hour counts.

Most of our growers who have made an effort and saved their crop, even though they had to flood their ground to get their beets up, are feeling pretty good these days. The weather is favorable for harvest, the roads are good, tares are low, and the yields in nearly all cases are better than expected. In fact, beets and alfalfa are the only crops worth mentioning in the Yellowstone and Clarks Fork Valleys this year.

Landlords who insist on cash rent and rent only for one year at a time are doing an injustice to themselves and their farms, for no one, under such a system, is interested in the future fertility of the land—the tenant being a one-year man, will take everything possible off the land without putting anything back, and the rental value of the land consequently is lowered from year to year.

JOHN MAIER, Agricultural Superintendent.

Loveland

Our growers are taking advantage of the good weather by speeding up their deliveries. More beets were received at the stations the first week than ever before.

Some very satisfactory yields are being reported. Such yields as 14 to 18 tons sound very good considering the grief we had to go through in producing a crop this year. However, the average yield wont be as high as last year. Those fields which were badly stripped by web worms are yielding only 5 to 6 tons, some even less. A number of our farmers who did not believe in spraying have now changed their minds as to its value. Fields which were badly infested with worms but were sprayed are yielding from 12 to 16 tons per acre.

We are glad to note that more of our growers are siloing their beet tops this year. With the present scarcity of pulp, due to every grower applying for his pulp, and the high prices of other feed, it behooves a farmer to silo his tops. By siloing, one can get 5 to 6 tons of cured silage per acre. With corn silage selling at $10 to $12 per ton, beet tops, when properly siloed, are worth from $50 to $60 per acre.

Most any night now it may be cold enough to freeze the beets which are pulled and topped. As you all know frozen beets are a source of a great deal of trouble both to the factory and to the farmer. We suggest that you insist on your beet labor covering the beets with tops. Your labor agrees to do this when he signs the contract.

We have found in a number of cases where the tares were high, the labor was not knocking the beets together before piling. See that your labor tries to get as much dirt off as possible. It is expensive for you to shovel and haul dirt, besides high tares at a station always cause more or less trouble. Also visit the tare houses and see how it is done.

We have had more or less trouble every year with rocks which come in with beets. Care should be taken in gravelly soil to see that all rocks are cleaned from place on which beets are to be piled by toppers. One rock can cost us several hundred dollars in knives, and a great waste of time. Half an hours delay on account of a rock getting into the slicer means two cars less at the dumps, and when eight or ten a day get in it makes it necessary for a lot more piling. Try your best to keep rocks out.

H. SCILLEY, Manager.

Time to Pick Brood Sows

Early fall is the time of year when hog growers should select from the spring crop of pigs the sows to be used as next year's dams. From now on until the breeding season these sows should be fed in a manner different from the way they were to be fattened for market purposes. Good forage crops are practically a necessity. A self-feeder containing a good quality of thrashed oats makes good feed for these sows. In addition, they may be fed a small amount of corn and some shorts or middlings, and fish meal or tankage.

Careful attention should be given to the selection of sows for breeding purposes. First, they should be selected from a good sized litter and from a litter whose dam has good suckling qualities. The brood sow should be of a good rangy type, with a good, strong arched back, deep sides, rather thin neck, and not too broad in the face. By all means she should be a "good-footed" animal with good, strong legs and pasterns and rather upstanding from the ground.

The sow that takes plenty of exercise freely is the one that is most likely to make a desirable brood sow. Do not select a sow from a litter whose dam is cross and irritable. The sow should be gentle and easily handled.

Attention should also be paid to the eyes. A blind sow is likely to step on and injure her pigs. Careful attention to these details at this time will have a tendency to add to the value of the breeding herd and to the number of pigs that may be brought to maturity.

Notes

H. Mendelson

Trash in Beets

In several districts the beets delivered at the dumps contain an unusual amount of beet tops and weeds. All factories are equipped to remove a normal amount of these by mechanical means before the beets enter the slicer. The abnormal amount cannot be handled. It gets in the slicer and clogs up the knives. This necessitates a frequent change of knives. If this meant only that it costs the Sugar Company a little more money to make sugar, the farmers would be little interested. However, every ton of beets delivered at the dumps and not sliced during the delivery season has to be shoveled off in the pile, which is done by the farmer. In one of our factories we could slice at least 100 tons of beets more per day if the beets contained only the normal amount of trash. Therefore, at the dumps delivering to this factory every day, 100 tons of beets, or about 30 loads, have to be unloaded into the pile which could be handled over the dump if the beets did not contain an unusual amount of trash.

Obviously it is to your interest to have your beets and your neighbors' beets as clean as possible.

This trash consists partly of beet tops used to protect the piles from freezing. The tops are worth more for feed than as obstructions in the factory. Part of the trash consists of pigweed and other weeds, which were more apparent in our fields at harvest time than ever before. Most of these weeds appeared after the help had finished what is optimistically called "third hoeing."

Perhaps another year we can prevail upon the help to perform their work in better agreement with the spirit if not the letter of the contract.

In some cases farmers explain the presence of trash and stones in their beets by the fact that the labor does not clean the ground on which the topped beets are thrown. A great many farmers do this themselves with a "V," doing in this way much better work in less time than the labor could do by fork or rake. Both parties are better satisfied under this method.

At Loveland they found the other day entering the slicer a beet which had grown through a horse shoe. The horse shoe, including nails, stuck to the beet. The beet was carefully topped, the topper evidently using great care to pare the top without hitting the horse shoe with his knife. It is hard to believe that the topper did not see the horse shoe.

Speedy Deliveries

By the morning of the 19th of October our Colorado farmers had delivered a little more than 700,000 tons of beets. Last year by this time the deliveries were about 460,000 tons, and at the corresponding date of 1917 they were about 650,000 tons. This is the speediest harvest we have ever experienced. A number of farmers will finish before October 25th. On the whole, men and horses are working at a higher speed than ever before.

In Kansas the farmers will reduce their wheat acreage by three million acres because they can't get cars enough to ship the wheat. Perhaps some potato farmers would like to ship their crop once at the same speed as the beet crop is taken off their hands. There have been a few delays here and there on account of railroad service, but on the whole we get as good service as can be reasonably expected. Very few farmers realize what an organization it takes to render this service.

We are glad our farmers are able to handle the crop in such a short time, but we have to state again and again that beets received in excess of the slicing capacity and the storage capacity at the factories have to be piled. The faster the delivery the more beets have to be piled. So at the end of the delivery season please do not complain too much of the delay at the dumps due to the necessity of piling beets. There is no other farm crop received by the buyer in such a short time with so few avoidable delays as the beet crop.

The Traffic at the Dumps

There is quite a variation in the way the farmers prefer to haul their beets. During the week ending October 19th the farmers at Nelson, near Brush, delivered the largest number of wagon loads in any one Colorado district, namely, 1,742, or an average of about 290 per day, although in this average is included one rainy day with only 197 loads. The maximum was 312 loads per day.

The largest number of loads delivered in the whole territory during this week was at Station 3, near Bayard, with 2,036 loads. The largest number of tons delivered was at Minatare, near Scottsbluff, where during this week about 6,700 tons were delivered in 1,922 loads.

There were at Nelson, with the exception of the rainy day, from ten to fifteen loads standing when the dump opened in the morning. At Minatare practically none was waiting, while at Station 3 from 12 to 30 were waiting. At a number of Colorado stations where we receive less than half the number of loads per day, often 20 to 30 loads are lined up before the dump opens. Why?

We gauge the efficiency of the dumps somewhat by the speed with which the early morning congestion is relieved, that is, by the number of empty wagons weighed back by 8 o'clock.

The figures for the week ending October 19th for the three stations mentioned are as follows:

	NELSON		MINATARE		STATION 3	
	Total loads received	Empty wagons weighed by 8 a. m.	Total loads received	Empty wagons weighed by 8 a. m.	Total loads received	Empty wagons weighed by 8 a. m.
Monday	305	41	356	56	439	47
Tuesday	311	40	315	81	314	38
Wednesday	197	17	165	2	176	16
Thursday	310	45	361	55	375	34
Friday	312	36	357	49	377	69
Saturday	307	46	368	49	355	73
Total for week....	1,742	225	1,922	292	2,036	277

At every one of these stations with extraordinarily heavy delivery the wagons present at the opening the dump had left the dump considerably before 8 a. m., although some loads were piled.

Beets Left in the Field

In quite a number of fields all over our districts a very perceptible number of beets escaped the puller and are left in the fields and nicely growing at the present time. It certainly seems worth while for a farmer with such fields at least to investigate whether this is due to a deficient beet puller or due to negligence, and correct another year whatever is wrong. In some fields it looks as if a man could make good day wages by digging this valuable raw material by fork or spade.

Trucks

There are many more trucks of all kinds used for the delivery of beets than was anticipated. We hope that farmers using trucks keep track of the real expense of delivering the crop, so their brother farmers may be able to utilize this experience.

There is one item of expense generally not included in cost figures by automobile, truck and tractor owners, and that is depreciation, although this is usually a very heavy item of expense. Nobody knows today how many years a truck will last used for hauling three-ton loads of beets over bad roads. It is almost certain that a new truck after such a season's work could not be sold for 75 per cent of its original price. The whole truck probably will be used up after four seasons. Therefore, a depreciation charge of at least 25 per cent per year is probably justified. If you use a $2,000 truck this means $500 per year. If you use the truck 150 days for all kinds of work, which is probably more than the average farmer will do, the item of depreciation is about $3.30 per day; or, if you haul 12 tons of beets per day, 26 cents per ton of beets.

Getting What You Pay For

F. G. Swoboda
State Chairman Pure-Bred Bull Campaign. Wisconsin

A farmer pays for a pure bred bull whether he owns one or not. The man who doesn't own one because he says he can't afford it pays for it just the same. It's easy to prove.

In a Langlade county herd the daughter of a pure-bred bull and a grade cow produced 338 pounds of butter fat in a year. In the same herd the daughter of a scrub bull and a grade cow produced 139 pounds of butterfat in a year. Loss due to scrub bull, 139 pounds of butter fat at 75 cents a pound, $104.25. That is his price for a pure bred.

In Brown county five herds with pure-bred sires averaged 85 pounds of butterfat per cow more than seven herds with grade or scrub sires. At 75 cents per pound the difference would be $63..75 on each cow; for ten cows, $637.50, the price of three pure-bred sires. The scrub herd owners paid for a pure bred three times over each year, but they didn't own one.

A Shawano county farmer increased his herd average from 2,214 pounds a cow, in 1911, to 8,099 pounds in 1918—a gain of 5,885 pounds a cow from using better sires. At $3 a hundred the increased yield was worth $176.55 per cow, the price of a good pure-bred sire. Many of his neighbors are still using scrub bulls. They are paying for the good bulls in their losses. Thousands of like cases may be given.

"Going, far beyond our fondest expectations." That's how A. W. Hopkins, Secretary of the Wisconsin Live Stock Breeders' Association, characterizes the pure-bred bull campaign. From every corner of the state reports come telling of new counties getting into the game. It begins to look as if a county had to get into the game or be hopelessly left behind in its live stock improvement.

No matter if you have only a few cows you can afford to use a high-class pure-bred bull. Four or five neighbors can "chip in" and buy a good sire. Use this bull for two years, then trade him for another of the same breed.

Danger in Rapid Rise of Land Prices

Prices of farming lands in many sections of the United States, especially in Iowa and the other Corn Belt States, have risen with such rapidity that serious consequences may be expected, especially with the return of normal conditions. This is a warning to farmers issued by the United States Department of Agriculture.

Speculators, many of them business men of the cities, in many cases have reaped big profits by buying and reselling without making any substantial contribution in return. Farmers have sold land at prices that seemed high and then have bought again at still higher figures, losing thousands of dollars in the exchange. The bona fide farmer who purchases land at present high prices may find the returns on his investment abnormally small if earnings should decline when normal conditions are restored, and may find himself seriously embarrassed if he has financed the purchase largely on credit.

That is a summary of the situation as it is viewed by investigators of the Bureau of Farm Management sent by the department into the regions of most marked speculative activity. Here is a summary of their advice which is broadly applicable probably to the majority of farmers in the regions affected, though it may not fit every individual case:

Much of the present speculative activity has been due to the fact that speculators have been able to catch some farmers unawares, buying their farms at a much lower figure than was justified in relation to prices in adjoining regions. Therefore you should be careful about selling to a speculator who is buying to sell again. If it is worth more to him it probably may be worth more to you. If you own a farm and desire to continue farming be very sure before selling that you can obtain out of the proceeds of your sale as good a farm as you have sold.

If you desire to purchase a farm you should be very sure that the price you pay is fully justified by the probable net earnings of the farm when conditions become more normal.

Be careful that the buyer of your farm is able to complete his payment on March 1 in case he fails to resell. In general it is safe to require an initial cash payment of one-third the sale price.

Be cautious about speculating yourself. Especially do not buy on a narrow cash margin with the expectation that you will be able to sell and obtain the necessary means of settling your contract. You may not be able to effect a sale. A land "boom" may collapse suddenly.—Weekly News Letter, U. S. Dept. of Agriculture.

Fig. 1.—WHITE'S EIGHT-HORSE MULTIPLE HITCH

Six Lead Horses Owned by Fred Muhme; Wheel Team Owned by Great Western Sugar Co.

540

Facts About Multiple Horse Hitches

P. H. McMaster

THE multiple hitch here shown was designed by Dr. E. A. White, Illinois Experiment Station, in co-operation with the Percheron Society of America and the Oliver Plow Company.

It was designed that more economical use might be made of farm horse power. The tandem hitches which are in general use in northern Colorado does not distribute the load equally on all teams.

The eight-horse hitch, shown on opposite page (Fig. I), works as follows: The lead team works against second swing team. Threaded through the pulley (Fig. II) behind the second swing team

Fig. II—Single Pulley for Second and Third Swing Teams

is a 2-foot chain. To one end of this chain is hitched the second swing team, and to the other end a 11-foot draw rod, to which is hitched the lead team. The lead and second swing team work against the first swing team. The pull is equalized by the block and tackle (Fig. III) placed immediately behind first swing team. This block and tackle consists of one movable and one fixed pulley. To the movable pulley is attached a 11-foot draw rod leading ahead to pulley behind second swing team. A chain 3 feet, 6 inches long is threaded through the movable and fixed pulley. The first swing team is attached to the free end of this chain. The wheel team works against the teams ahead of it. The pull is equalized by means of a block and tackle, consisting of a movable, single pulley block and a stationary double pulley block. The wheel team is attached to the free.end of the chain with which the block is threaded; a draw-rod 11 feet long leading ahead connects with pulley ahead.

The load is so equalized by this system of blocks and tackles that each team draws the same load.

The angle of pull is at right angles to the shoulders of the horse. It is the angle of trace rather than the distance from the load which determines the horse power needed to move a given load. The weight of the blocks in the first three teams and a 50-pound weight attached to draw-rod behind lead team makes the angle of trace true and does not increase draft.

One man and eight horses can pull a gang of three 14-inch bottoms and plow 8 to 8½ acres per day. You save one man and plow by using one eight-horse team instead of two four-horse outfits.

You handle only one pair lines, which are on lead team. The

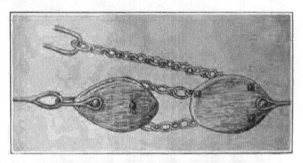

Fig. III—Movable and Stationary Pulley

other teams are "tied in" and "bucked back," giving automatic control over them.

These hitches may be used on your binder if equipped with truck tongue.

Horses have freedom to move in any direction, thus eliminating the fretting sometimes occasioned by other hitches.

The farm horse has not been given proper consideration in attempting to compare the relative merits of motor power for the farm and horse power. Perhaps we will find the horse the most economical source of power for small farms if efficient methods are adopted in utilizing this power.

The Illinois Experiment Station through a very comprehensive study of detailed costs found that the horse is the most economical and efficient source of farm power on farms under 260 acres.

The cost of these hitches seem almost prohibitive at first sight, but when the horse power, efficiency and the welfare of your horses is considered, they are well worth the money.

4-horse hitch	$22.50
5-horse hitch	27.50
6-horse hitch	44.75
8-horse hitch	68.50

The White multiple horse hitches can be secured from your local county agents. If an order for five or more is sent in at one time there is a discount of 5 per cent allowed on the above list price. H. H. Simpson, county agent, Boulder County, Longmont, Colo., can furnish these hitches to anyone desiring same.

"To Plow or Not to Plow"

H. H. Griffin

IN the May, 1919, issue of Through the Leaves, Mr. J. D. Pancake asks the above ·question, after reciting his experience with a crop of beets in 1918, on a field some of which had been fall-plowed and some spring-plowed. From this field of about 17 acres he estimated his loss from fall-plowing to be at least $300. I would suggest to those having this number of acres that it will be well to refer to this article. Both the editor and Mr. Pancake ask for an explanation of the seemingly peculiar results that he secured and I have been anticipating, as well as hoping, for some months that some one would make reply to this query. This is of great importance and demands careful thought for there is no manipulation in farm practice more potent for good or evil than is the preparation of the soil, which is virtually plowing, for if a field plows well there is not much left in the after preparation.

Now if Mr. Pancake were the only one who has had experience similar to this we might dismiss the subject as of little importance and forget about it, but the writer has come in contact with a number of such examples in the past few years which have made him "sit up and take notice." The chances are that many others would have done the same had they had fields where comparisons could be made. The fortunate thing with Mr. Pancake is that he had a condition where a logical comparison could be made, otherwise the results he secured would not have been apparent and he would probably have gone on losing money and been none the wiser. It does not hurt anyone to lose if he doesn't know it.

Professor Bailey, of New York, once went before his class exhibiting in one hand some soil in a nice, moist, friable condition; in the other hand he had some of the same soil which was a mass of clods. He was making this comparison to impress upon his students that these two soils were actually the same as regards composition and fertility, but wholly dissimilar in productive capacity, due to the difference in physical composition.

Plowing is the manipulation on the farm that we should depend upon to put the soil in proper physical condition. If the soil plows poorly, turning up cloddy, no amount of after work can get as good tilth or production as could have been secured had the land plowed properly. Nor should a soil, in plowing, turn over a loose, dry, lifeless mass. When we say "properly plowed," what do we mean? We mean that when the soil, while plowing, is pushed up against the mold board and is turned, it has that amount of moisture that will crumble it into a fine granular condition, in other words, turning it over into a mass so composed that the rootlets of the small plant can find the greatest amount of moist particles as a feeding ground. This vast numbr of minute particles will, of course, absorb and retain the maximum amount of moisture, and further this packing and crumbling together with the moisture will place it in such a condition that the sun and air can work those chemical changes that make for production. A soil of this character is also in a condition to be given just that proper firming conducive to the best plant growth. It is difficult to express on paper just what proper plowing should be, but most all farmers are aware of what it should comprise, and it stands to reason that the time to do this is when it is most likely these results will be procured, be that in the fall or in the spring.

We are well aware that in farming it is not always possible, especially in plowing, to do it just at the right time, for there is a vast difference in soils, with which moisture conditions play a great part. It is difficult with some soils to get a moisture condition just right, it is either so wet that when it is turned it becomes glazed, or it is so dry as to turn very cloddy. If spring plowed, the former condition is the more detrimental. For many reasons the statement "fall plow" or "spring plow" may not be conducive to the best results. The farmer should understand what constitutes good plowing and then by coupling this with the knowledge of his soil, the crop that he intends to grow, and the probable weather conditions with which he will have to contend, act when the plowing can be done in the best possible manner.

The East can set no precedent for time of plowing in Colorado, any more than it can for methods of fertilizing, manner of harvesting, maintenance of roads, etc. Our climatic, as well as soil, conditions compel us to have methods peculiarly our own. For instance, we have what is termed "gumbo" soil, a class of soil that is seldom in fit condition to plow. It either has so much moisture that it will not slip from the mold board properly or else turns over very cloddy. It is difficult to find it in just the right condition for plowing. It is imperative to fall-plow such lands, for the winter frost action must be had to give it tilth in either case. Mr. John Maier expresses it correctly in the September issue when he says "pour a little water over gumbo and you get about the same "slacking" results. as if it were poured over lime." It is seldom that winter moisture is so scant that you cannot get the action of which he speaks. The farmer is compelled, to get satisfactory results, to fall-plow this kind of land. Such soils, after being acted upon by the winter frost, blow badly and the proper methods must be used to minimize this.

Our dry winters and springs, coupled with the winds, put a different aspect upon the matter of plowing from that in the East. Were we to be sure of plenty of winter moisture there is no question but that fall-plowing every acre would be the proper thing to do. We must take these things as they are in this arid country. There are some sandy soils in northern Colorado that it is the wrong thing to do to plow in the fall on account of wind action and loss of moisture. This is probably true four-fifths of the time. Such soils never become so moist that they cannot be properly plowed at any time in the spring, then it is a losing game to plow it up in the fall to be shifted by the winds for some months and dry out the surface for a number of inches. Furthermore, plowing such soils in the dry time puts a mulch over the moisture below the plow slice and prevents it rising to support the young plant. Planting and germination are not hastened. Generally if such soils are spring-plowed and seeded at once the moisture content is uniform throughout and the young plant gets an even start. What is true of moisture content as affecting germination in the soil above mentioned is also usually true on our soil that may be termed "clay soil." Soils that we would not class as "potato soil," yet not so heavy but what we may work them at any time of year without injurious results so far as moisture content is concerned. These are the soils to which the injunction "plow properly" more aptly applies. Such soils, if plowed in a cloddy condition in the fall, and the winter remains dry, will not pulverize as the gumbo, but remain a great mass of inert clods. The action of the winter frost seems to have no beneficial effect. The moisture is dried out deeply because the winds have an opportunity to penetrate to the bottom of the furrow slice, often drying the soil below that portion plowed. It requires a very heavy spring storm to reach the subsoil and put the land in productive condition. Should such soil plow fairly well in the fall, and the weather remains dry until late in the spring, the young beets cannot germinate well and get the right hold and make the crop that they would were the planting done when the moisture was more uniform throughout the plowed mass. Proper firming is difficult with such soil plowed in the fall, while the moisture content in spring generally allows of proper firming. The writer has in mind two instances, as above cited, on some of the best soils in the Fort Collins district where this year the crops on the fall-plowed land are at least one-quarter less than on the spring-plowed. These fields are side by side and the comparison easily made. The farmer often can afford to plow such soil in the fall, if the plowing can be done in a somewhat favorable manner, and take the chance of the future moisture content being favorable, for there are compensating features, such as getting the work well under way; getting a little better depth of plowing, due to frost action on the sub-soil turned to the surface, etc. The fall-plowing on any soil other than gumbo is better for any other crop than for beets, which can be explained partly on account of the moisture content and partly from the fact that most any other crop put on the land plowed at this time does not penetrate the soil so deeply, and hence moisture content does not affect them in the same manner. If the clay soil turns up cloddy and the winter remains dry a good seed bed is impossible and it is doubtful if spring storms will be sufficiently heavy to wet the soil so that the beets will continue to thrive, or if the plowed portion becomes wet often the sub-soil will remain dry and the growth be retarded, just as soon as the root reaches this rather dry earth; only very early irrigation will relieve it.

I have shown some causes contributing to poor results to beets on fall-plowed land that have come to my notice under conditions where comparisons could be made, as was the case with Mr. Pancake. To say that his results should be attributed to causes herein mentioned is impossible without knowing more of the conditions affecting the soil and moisture than he mentions, but those I have given are to my mind the reasons that gave similar results on soils much the same as that he is farming, and I believe he can see in one or more of these causes the reasons for the results he obtained.

H. H. GRIFFIN.

Representative Livestock Sales at Denver Yards

The "Record Stockman" publishes each day a list of "representative sales," showing the number of animals, weight and price. The following tabulation is the total number of head reported in such sales and is intended to show the trend of prices for the first twenty days of October:

FEEDERS

	$9.00 to $9.45	$9.50 to $9.95	$10.00 to $10.45	$10.50 to $10.95	$11.00 to $11.45	$11.50 and Up
800 to 899 lbs...806	1,223	935	528	86	......
900 to 999 lbs...607	444	886	1,226	64	105
1,000 to 1,099 lbs...140	273	532	248	282	21

800-lb. feeders of good quality sold from $9.75 to $10.75, averaging about $10.25.

900-lb. stock of good quality sold from $10.00 to $10.25, averaging about $10.50.

1,000-lb. stock of good quality sold from $10.00 to $10.50, averaging about $10.25.

Choice stock brought a little higher prices, some 1,000-lb. stock bringing $14.25. Likewise poorer qualities brought considerable less than average prices.

LAMBS

	$11.50 to $11.95	$12.00 to $12.45	$13.00 to $13.45	$13.50 to $13.95	$14.00 to $14.45	$14.50 to $14.95
50 to 59 lbs..... 45	148	2,879	2,457	1,489	......
60 to 69 lbs.....2,025	828	4,235	1,374	214	268
70 to 79 lbs...... 145	......	256	576	265	2,226

Feeder lambs have been in demand by both Western and Eastern buyers; $13.00 to $13.50 for desirable stock was the ruling prices.

EWES

	$5.50 to $5.95	$6.00 to $6.45	$6.50 to $6.95	$7.00 to $7.45	$7.50 to $7.95	$8.00 to $8.45
90 to 99 lbs.........1,883	447	266	1,720	375	1,498

The ruling prices for ewes was from $6.00 to $7.00 for the period for desirable stock.

547

A Beet Top Silo. Tops from 30 Acres are Herein Siloed. Farm of Denzel Hartshorn

Showing Simple Device which Aids in Unloading Tops at the Silo

Handy Appliance for Unloading Beet Tops

Alfred R. Williams

The progressive farmer is always on the lookout for anything that will help him to do his work easier and with a saving of time; therefore, we believe that our growers will be interested in the accompanying cuts, showing a quickly made device which is a help in unloading beet tops, particularly if the tops are unloaded into a pit silo.

As will be seen, this device is nothing more than an inverted V, or trough, about two feet high. Narrow strips are nailed with small

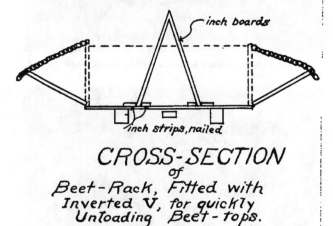

inch boards

inch strips, nailed

CROSS-SECTION
of
Beet - Rack, Fitted with
Inverted V, for quickly
Unloading Beet - tops.

nails to the beet bed as shown, and the V toe-nailed to the strips. The strips also help to keep the V steady.

There is really no weight or strain to come on this device, as its purpose is to throw the weight of the tops outward toward the sides, and with the tops loaded on the beet bed with the sides at about the slope shown, they readily slide off when the end chains are loosened, letting the sideboards down. The bulk of the tops rest on the sloping sideboards.

Mr. Denzel Hartshorn, one of our growers, has used this device with much success in hauling his beet tops and states that it is a great time and labor saver in unloading.

Rambouillet Ram Sold by S. R. Burton to F. S. King Bros. Co. at $400

Hampshire Ram Sold by R. S. Blastock to Robt. Taylor at $200

The National Western Ram Sale

J. R. Wood

The second annual ram sale, held during the first days of October in the stadium at the Denver Stock Yards, closed with much satisfaction to the promoters.

The promoters are the same group of men who put forth so much effort in the annual live stock show held during each winter at the stadium, and both shows or sales are instituted for the same purpose—the stocking of blooded animals on our western farms to make more profitable this farm product, in place of raising any old "scrub" stock that might be at hand, at little or no profit.

It is conceded by everyone, that blooded stock does not predominate on our western farms and to better conditions in sheep raising and wool production the ram sale was inaugurated both, that better stock might be produced and that more might become interested in sheep raising.

All of the famous strains of sheep were represented—the Rambouillet, Hampshire, Cotswold, Shropshire and Lincoln, coming from Montana, Idaho, Kentucky and some from Colorado.

Prices in general were a little lower than last year's, but a long list of sales indicates the many buyers present who were ready when real bargains were offered. The following tabulation gives an idea as to the range of prices and the number sold at such prices, and are representative of the entire sale:

	Under $50	$50 to $74	$75 to $99	$100 to $149	$150 to $199	$200 to $250	$275	$330 to $350	$400
Rambouillet	40	170	48	27	10	3	1	3	1
Hampshire	211	28	6	13	11	4	0	1	0
Shropshire	33	12	1	0	0	0	0	0	0
Delaine	18	4	1	0	0	0	0	0	0
Lincoln	7	10	6	0	0	0	0	0	0
Corriedale	46	11	15	0	0	0	0	0	0
Cotswold	0	1	1	4	1	0	0	0	0

The Rambouillet and Delaine are designated as "wool" sheep and Corriedale, Hampshire, Shropshire and Lincoln are "mutton" sheep.

Our western farms should be better stocked, not only in numbers, but in quality, and sales of this nature not only afford an opportunity to buy good stock, but also afford an opportunity to become acquainted with others in the same line of business and an exchange of ideas is usually very profitable.

The accompanying photo of the Rambouillet ram is one sold by S. R. Burton to F. S. King Bros. Co. at Laramie, Wyo., at $400. The Hampshire ram is one sold by R. S. Blastock to Robt. Taylor, of Abbott, Neb., at $200.

551

BEET TOP ENSILAGE IN THE MAKING

Has Sugar Beet Made Good Here?

George C. Rice

AS the sugar beet made good in western Montana? This is a question of vital importance to the residents of the fertile valleys in this section of the state. After two years of battling against the combined handicaps of a world beets. Part of them in fair condition, war with record prices for wheat and the natural hesitation on the part of the rancher to try out a new crop, the sugar beet has made a very favorable record this year in our territory.

Although the acreage for the Missoula factory is still small, the tonnage per acre will be an agreeable surprise to the growers. Many fields will run 16 to 18 tons to the acre. Over half of all the acreage from the big Missoula mill will yield 15 tons per acre if the growers have sufficient water to finish the crop. The tonnage will not be much greater than last year, but there will be reason for gratification for those that are interested in the industry in the fact that the attitude of many of the farmers has completely changed toward the industry. Instead of being in an antagonistic frame of mind now many of the farmers have expressed themselves as favorable to the growing of the beet because they have had proof that it is a beneficial and profitable crop to grow. Fields that have been overrun with weeds have been cleaned up, and the fertility of the soil increased by the intensive cultivation. The fields that yielded a crop of beets last year are this year yielding double the crop of grain that would have been possible before.

The contrast between the beet land and the other land is so pronounced that it is very noticeable. As a result of this condition it is hoped that the farmers will put in twice the acreage next year that is in beets this season.

In company with R. J. Dee, traveling freight and passenger agent of the Northern Pacific railroad, and R. M. Barr of the Great Western Sugar Company, I visited the beet fields in the Bitter Root valley and the following are some of our observations on the trip.

Our first stop was at Lolo, where we visited the farms of Tyler Worden, Dr. Mills, T. W. Lee, W. H. Rock, and other small fields and found the beets in excellent condition and with few exceptions they were well cultivated and absolutely free of weeds. At McClain Spur we found beets on the Daniels and Maclay farms also in good condition. All of these beets were irrigated from Lolo Creek in which the water is getting low and some fields are suffering by not getting the water to the crop fast enough. Under these conditions Mr. Barr states that he expects fair returns from those crops.

At Bass we found about 200 acres of beets. Part of them in fair condition, and part of them poor on account of being planted in land that had received little irrigation in the past three years.

The difference in crops where land was well irrigated to September of the past year was very noticeable, and we found on many farms where this had been done that the growers were giving the beets the first irrigation of the season and their crops were in good condition.

For Stevensville and Bing dumps we looked over 500 acres of good beets, most of them well cared for. We looked over the crops of Cerino, Irvine, Lane, Sanders, F. L. and E. E. Williamson, Taylor McKnight and G. B. Strange and were much impressed with the clean, well cultivated fields and the prospect of a good crop.

We examined closely the Great Western Sugar Company's operations here on the Warner farm, which they have rented, and on the G. B. Strange farm which they bought. On the two farms they have approximately 300 acres of good beets and the whole crop shows the effect of intensive farming. The prospects are that this crop will average about 15 tons per acre and Mr. Barr states that this is a good average on a large acreage, while individual patches on this acreage look good for 20 tons per acre.

From here we visited the Mendel and Wood farms that will send over 100 acres of good beets to Victor. The Mendel farm has a field of the best looking beets in the valley. Albert Wood has given the beet laborers an interest in the crop, and is well satisfied with his prospect and the fact that he is cleaning up a lot of very foul land.

We then visited the Humble farm, owned by the sugar company. On this farm they are growing 200 acres of beets that are in fine condition. This farm has been producing grain and alfalfa for years and Mr. Humble has rented portions of it for years previous to selling and I am informed that when the sugar company got it, it was one of the foulest farms in the valley. This is their second year under very bad labor conditions and I should say that in three or four years, with their system of farming, this will be one of the model farms in the Bitter Root valley.

In the fields in which they are growing beets this season they have eliminated beautiful crops of mustard, wild oats, Canadian thistles, quack grass and morning glory, and intend to clean up as much more land next season.

For the beet laborers they employ to do the hand work on these crops, they have built comfortable cottages and give each family a piece of land adjoining to grow garden truck. As these families are only working on the beets part of the time, they employ them for irrigating, haying, harvesting or any kind of farm work at whatever wage is being paid in the valley; thus having through the beet industry, a great asset in labor on the farm.

W. S. Bailey, a joining farmer, is also growing a fair crop of beets at this station.

From here we motored over the Woodside, Riverside and Hamilton beet fields and saw a lot of fine beets, but as it was getting late, with the exception of W. O. Logan's field, the others were viewed from the road. Logan's field north of Corvallis we visited and found one of the best cultivated fields in the valley, and one that Mr. Logan and the sugar company employes are justly proud of. He has followed instructions of fieldmen employed by the sugar company to the letter and the field and crop shows what can be done by intensive work. Other fields we viewed in this district were T. R. Glass, J. E. Dressel, J. H. Hawker, William Hay, A. W. Keays, Charles Goneau, Otto Quast, S. A. Burkhart, Slack Bros., M. F. Meador, Christoffersen Bros., W. F. Wehr and John Kalberer, and with few exceptions the crops looked good.

The few farmers we interviewed on the trip were well satisfied with the outlook and many of them intend to increase their beet acreage next season.

The hay, grain and pea crop throughout the valley also look very promising, but many of those fields are very foul and some of them so foul that we had to stop and study them to say what the crop was.

The conclusions we arrived at were that if the farmers close enough to the railroad were devoting a small part of their farm to sugar beets and rotating crops, that weeds would be eliminated and the land would be producing double what it is producing today.

Besides the profits from the commercial beet, Mr. Barr stated that where the beet tops were fed judiciously to dairy stock their feeding value had gone as high as $40 per acre in some instances.

Mr. Barr has promised me a trip through the Missoula, Grass Valley, Flathead and Flint Creek districts, where he states the beets are in much the same condition as in Bitter Root, and we hope to be able to write a little more, later on, about this crop which means so much to the farmers of western Montana.

Sheep, lambs and cattle are pastured on tops in all beet districts with good results, and following is an article from the Breeders Gazette by L. Ogilvey, an authority on beet top feeding.

An average acre of sugar beets produces two tons of beet tops and 1,200 pounds of dry pulp, or 7,000 pounds wet. Eighty per cent of the dry pulp is digestible, 50 per cent of the dry beet tops. Several methods of consuming beet tops are used. They are pastured from the field as topped or piled or hauled to the corrals and fed (about 17 to 20 pounds a head per day with other feeds), or ensiled, which breaks down the proteins into a more fattening feed and makes them easier of consumption without waste by the cattle. A great many more beet tops would be ensiled, rendering them less tough and leathery, were it not that it has to be done at the busiest time of beet harvest. When piled to dry, the piles should be small so as to avoid heating and either hauled or shocked as they dry. The tops should be well shaken to free them from dirt.

Sheep and lambs eat the tops best while still green, and waste as little in the field as when pulling the toughened tops through the racks and tramping them under foot. If stock is to be fattened while picking them from the field, there must be an unlimited quantity. Efforts to make the stock clean them up will result in losing the flesh put on when full fed. As a rule none will be wasted if other lighter fed stock follow. Often snow will fall before the stock can gather the tops and they will be tramped. For this reason they are often piled, though the stock gathers them. An acre will graze a steer 100 days, but beet tops are not a balanced ration and should be fed with other feeds. Notwithstanding this, the stock will make a great deal of weight on beet tops fed ad libitum for a short time, though it may be largely "fill." Nothing puts stock in better condition for the feed lot and gradual introduction to stronger feeding.

There is no sweeter beef than that made on beet tops. It contains too much water for shipping, but when locally killed it is much like the best Scots beef, and one can say no more as to its excellence. This is especially true when the cattle come from mountain pasture and take a short fill and revivifier on beet tops after their journey. The beet crop is assuming more importance to the beef maker as the importance of its by-product is realized. In these days of expensive feed it is essential that cattle be put in condition to make gains before the golden corn and diamond-priced cotton cake are poured to them. For this purpose beet silage, tops, and corn silage are eminently suitable. Apart from their feed value they are conditioners and enlarge the stomach of the western steer off short, dry grass in an easy manner, at the same time correcting the dry, feverish condition that often ensues on shipping. Combined with alfalfa, the condition of the stomach will be loose enough; in fact, the amounts of laxative feed may have to be regulated, but that is far preferable to trying to put on gains on dried-up cattle.

"I know when the farmer has adversity he always goes to the dairy cow for assistance and she never refuses. In prosperity she is forgotten. That is human nature. It is only the farsighted, conservative people that seem to fully realize her importance and her value."— W. W. Marple.

One Editor's Views

(Editors' Note: These editorials from Successful Farming are timely and of such special interest that we are reprinting.)

BALANCE WHEEL NEEDED

You know what happens when the balance wheel gets to slipping on the shaft. The thing to do is to drive the wedge pin in again, or tighten the set bolt. It is the only thing that will make the engine or machine run steady.

Throughout the world the balance wheels in the political and economic machinery of governments have been slipping of late. In some cases the balance wheels have flown off and smashed governments so that new machinery had to be set up. While the new machinery was being installed chaos has reigned.

In this country we find a condition that is alarming. There is a mingling of radical discontent with ultra conservatism in every line of thought and endeavor. The farmer element of our national life has always been relied upon to act as a balance wheel to the political machinery, or ballast to the ship of state. But now there is considerable unrest among farmers. If they were as well organized as city labor would they be as radical? Would they still be the balance wheel of the nation?

Back of every effect is a cause. When the social machine runs wild without a balance wheel, too little heed is given to the causes that distress the people. Radicals make a great noise and get the attention away from facts and causes. Interests are so interlocked that when "the party of the first part" says to "the party of the second part," go hang, both parties are going to suffer.

Clear thinking and cool action are needed now more than a hullabaloo from a hot headed agitator who has little to lose. Farmers must be relied upon more than ever before to give their conservative judgment in the councils of the state and nation. They must play their part now as never before in big affairs. Their organized voice must speak for action stripped of all political tommyrot—action that goes to the cause of our troubles.

PEACE OR PREPAREDNESS

We can rest assured that if the League of Nations covenant is rejected by our senate the next step is a big army and navy and compulsory military training for every young man. The league covenant will limit the armament of nations and take the manufacture of munitions of war out of private hands so far as possible. Johnson,

Borah, Knox and the rest have not stated these things.

We must either become a controlling influence among nations in a league which will reduce the cost of preparedness; will lessen the possibilities of trouble betwen nations; will work for harmony among the nations of the world, or we must build a larger navy, increase our army, and continue to levy the present or a greater burden of war taxes. Our choice lies between perpetual peace with a chance in a lifetime of having to send a few soldiers to help police an obstreperous nation, or militarizing this country and adopting compulsory military training in order to go it alone.

PAYING TEACHERS MORE

We are gratified to learn of so many school boards that have advanced the salaries of rural teachers to a point where the best teachers can be secured and the standard of education advanced in those communities. It has been a burning shame that so many teachers have been paid a less wage than that paid to herdsmen on the farms who take care of livestock. As if the calves and the pigs were of more value in the minds of the farmers than their children!

No, this is not so; has not been so, but the work of hiring teachers has been left to the school boards while the patrons of the schools, anxious as anybody to have good schools, have gone about their business of farming. School boards in rural districts are composed of busy farmers. It has only recently been forced upon their attention that school teachers can no longer be secured at the pre-war prices. The critical shortage of teachers compelled them to take notice of the pleas of Successful Farming and other publications for a higher salary for teachers.

The farmers have the money. They can and do ride in the best autos. They can and will have the best teachers, for they can afford them. Some school boards have not yet come to realize that these are changed times and must be met or the rural school houses will be closed for lack of teachers. We suggest to any district that has not succeeded in securing a teacher at the pre-war salary to try a bait of $100 a month and see how long the school will be without a teacher. Dignify education by making it worth having and worth paying for—then get the best teachers by paying them salaries in keeping with the times.

MUCH ADO ABOUT NOTHING

One thing many fear is that England shall have six votes to our one, or that there are say 32 votes against us, as Senator Reed of Missouri so tearfully bewails. What of it? What if there were three hundred votes against our one, the treaty reads that "Except where otherwise expressly provided in the covenant, decisions at any meeting of the assembly or of the council shall require the agreement of all the members of the league represented at the meeting." Unanimous vote is required except in four instances, one in reference to "procedure of meetings and appointment of committees to investigate particular matters;" an approval of the unanimous choice for secretary general of the league; when a dispute is referred to the assembly by unanimous action of the council, and when an amendment can be made by a majority of the assembly after a unanimous vote of the council. So, in none of these cases can any vital matter get by without the consent of the United States.

But let us view it from another angle. Suppose the British Empire had six votes to our one, can any American really believe that the English speaking nations would be a menace to our security Have they ever been? Is it not a thing most desirable that the British colonies of Canada, New Zealand, Australia and South Africa should be recognized before the world as self-governing peoples? Would anyone, to gain bogy security to the United States, have these great nations tied to the British Isles without independent sovereignty?

SUPPOSE FARMERS STRIKE

The farmers keep their heads when all others fly off the track. We ask city workers to ponder a moment what might happen if the farmers should do what the workers are doing—demanding shorter hours and higher pay. The farmers are their own bosses, so they would not have to quarrel with anybody. They could hold out on strike until they got good and ready, for they can feed themselves.

You working city fellows, suppose for a moment that the farmers adopted the eight-hour day. It would cut down production at least half. Suppose they also set a price on their labor and their products based on an eight-hour basic scale. Where would you get your food? Only the rich could buy it at all, for the price would be prohibitive to men on strike. If the cost of living is too high now, how will lessened production affect it? How will increased cost of porduction bring prices down? You live now because the farmers have gone on producing, working nearer sixteen hours a day than eight hours. You can buy food because the farmers have not gone on strike, have not ceased to produce, have not cornered the market and said "we demand so much for our products or we won't work."

If you city workers expect the farmers to go on feeding you at the old price, you have got to get back to work at the old wage and make it possible for the farmer to buy cheaper so he can produce cheaper. This is not a one-sided game. It takes two to play it, and if you city fellows quit, don't get sore if you go hungry soon. Either the farmers must do as you are doing, shorten the hours and demand higher pay, or else you must lengthen the hours and produce more without more pay.

The farmers have been patient with you. When they lose their patience, look out. You have already taken their help. If they quit, too, who is going to feed you. What city workers have in common with farmers is not so much political as economic. What are you going to do about it?

IN WAR—IN PEACE

It was a great event when the United War Work campaign of last fall raised over $200,000,000 from the American people of all creeds, races, parties and stations in life. It was the largest voluntary offering in all history. The sum is staggering. It showed what we were willing to do as a patriotic duty.

Did it ever occur to you what such a sum might do in times of peace when applied to relief work or social betterment? Are the social diseases a less menace in peace than in war? Are the sufferings of poor widows and orphans any less in peace than in war? Does the sting of cold when fuel cannot be bought, bite any deeper in war time than in peace time? Will we now forget the soldier whom we idolized a few months ago?

Two hundred million dollars could remedy tenement congestion, rout disease, put those in good hospitals who cannot even call a police physician, rid our land of many a dark, vile hole of iniquity and make this country a place where Bolshevism could not gain a foothold.

But back of that $200,000,000 was the willingness to give and sacrifice. This spirit carried into peace times will make this country great. But alas, the war is over, greed comes boldly forth, and next will come the scandals of after the war. Patriotism is so short lived, so often mere veneer.

BLOWING BUBBLES

If a shoe pinches and hurts a corn it does no good to blame the shoemaker. The fault lies in selecting the wrong size to wear. If one gets a briar in his flesh, no one but a fool would press it further in. If a tyrant is overthrown by a revolution, it is no relief to place another tyrant on the throne.

Humanity seems to be just as foolish as that. War obliterated hundreds of thousands of tons of coal, of iron and steel, of food stuffs, of clothing, and killed millions of men who had been producers. War turned factories of peace into factories of war. Food was scarce, everything upset and out of harmony with peace times. Millions of men were fed, clothed and cared for by their governments. Suddenly these abnormal conditions ceased. Once more the thoughts of a world were of home and peace. But factories cannot as readily be transformed again into peace factories. It takes time to again find a market for peace goods. Nations and individuals are hard up—nearly bankrupt. Manufacturers need cash with which to pay labor and buy materials. Cash is hard to get and worth about half pre-war values.

Radicals took advantage of the situation and poisoned the minds of millions of workers. They taught that increased wages and shorter hours were the panacea for the higher cost of living. They blew beautiful bubbles into the air and millions of people gazed upon their beauty and reached with childish glee for their possession.

They grasp the idea of less work at higher wages only to find that flooding coal mines, destroying factories, lessening production and demanding higher wages only increase the cost of living so that no real advantage has been gained. The beautiful bubbles break when so roughly handled by impractical men. Labor leaders who teach that they can increase wages and reduce the hours of labor without increasing the cost of living are only blowing bubbles. Farmers who think they can prosper more when by accident or design the output of farm products is lessened and prices go soaring, are only gazing at beautiful bubbles.

Nothing will be gained by unseating the so-called tyrant capitalist employer and by putting the tyrant radical labor leader in his place. Nothing will be gained by placing the country in the hands of a labor or producers' monopoly if they have no more heart and conscience than capital monopoly.

Colorado as a Dairy State

Roy M. Curtis
Tester, Johnstown Cow Testing Association

OLORADO as a dairy state has many advantages over other states and it is one of the coming dairy states. We have all of the feeds most suitable to dairy cows and we have a climate for the dairy cow that cannot be excelled, which means far more than many people realize. The climate is not only healthy, but the winters are warm enough that the cow does not have to be housed where they are stanchioned with nothing but a cold cement floor to sleep on and no chance to get a drink when she wants it, and I find that the dairy cattle in Colorado drink almost as much water at night as they do through the day. In the colder climates

561

they do not drink but once a day in the winter time, and the result is a decrease in the flow of milk.

Our dairy cattle do better when they have comfortable sheds partly open to the south and plenty of straw to lie on. The air is then pure and she has a chance to get a drink when she wants it, her bed is more comfortable and she has more chance to keep clean. She is driven into the dairy barn the next morning to be milked. The barn is clean and the air is fresh, for there hasn't been any stock in the barn over night.

Colorado is very new in the dairy business, but some of the dairy men are just beginning to show what can be done. Lohry & Wailes of Johnstown have one cow that without any special care has produced in nine months over nine tons of milk, the exact figures being 18,200 pounds and 728 pounds of butter and is still milking 50 pounds of milk per day, testing 3.8 per cent, while J. W. Whowell of Johnstown has a cow that in eight months has produced 719 pounds of butter from 18,100 pounds of milk and is still making 65 pounds of milk per day, and has been fresh since February. I could mention other big records and we expect to make still larger in the future. These men are in the Johnstown Cow Testing Association and the cow's rations are made out each month by the tester.

Colorado has many cheap feeds that give us the variety and the kinds most suitable to the dairy cows. Besides Colorado being a great alfalfa state, which is very necessary for a dairy country, it has its beet tops, dried beet pulp and syrup, all of which are very good and cheap milk producers, and feeds that you will find in every ration of the high producing cows of the Johnstown Cow Testing Association. Beet tops are a very rich and cheap feed that gives us a variety that can't be found in other feeds; dried beet pulp, which is also one of our cheap feeds, gives us a variety which is very important in our rations, and also takes the place of some of our carbohydrate feeds. Dried beet pulp is also an excellent feed for growing cattle and for calves. Syrup is a very good stimulant upon the milk flow and appetite and is a great conditioner. It also gives our ration more variety.

As we advance and learn more about the dairy business the more profitable it becomes. Many of our Colorado dairy men are very new at the business and expect their cows to produce without sufficient food. The Testing Association is helping the dairy men to get started and in some cases the profits have been doubled.

In the C. H. McNeil herd the butter fat was increased in one year from 234 pounds to 326 pounds per cow per year, while the profit above feed was increased from $56 to $102 per cow per year.

Many of our Colorado dairy men are putting their ideals high, which is the way they should, and say they are going to have as good dairy cattle as can be found anywhere in the United States.

PASTURE GRASSES

(1) Brome grass. (2) Orchard grass. (3) Meadow fescue. (4) Timothy. (5) Kentucky blue grass. (6) Tall meadow oat grass. (7) Redtop. (8) Western wheat grass. (9) Slender wheat grass. Canada blue grass and wood meadow grass are the only other perennial grasses which have any importance for seeding permanent pastures in Nebraska.

Pasture Mixture for Thin Sandy Soils in Eastern Nebraska.

Crops	Seed per acre	Weight per bushel	Estimated price per bushel	Price per pound at bushel rate	Cost of seed for one acre
	Pounds	Pounds	Dollars	Dollars	Dollars
Redtop	6	14	3.00	.214	1.28
Canada blue grass	3	14	2.00	.143	.43
Tall meadow oat grass	3	14	4.00	.286	.86
Brome grass	5	14	2.50	.179	.89
Alfalfa or sweet clover	3	60	12.00	.200	.60

Total cost of mixture per acre, $4.06.

Care of the Brood Sow from Breeding Season until Weaning of Pigs

A. J. Lovejoy

TO have the brood sows in prime condition at the beginning of the breeding season, in the fall, it is well to have them come off good fresh grass or pasture of some kind, having for a short time been fed grain and being in a slightly gaining condition. This usually brings them in season very shortly, and generally where there are many, all will come in season about the same time. This is well, as it enables a large number to be bred as near the same time as possible, and the litters to arrive about the same time in the spring, thereby giving one a large number of youngsters of practically the same age.

During the time these sows are being bred and carrying their litters they should be fed absolutely right for best results. First and all the time they should have plenty of exercise, the more the better. If they could run about the pastures and fields during the daytime, between the feeding periods, it would be well, and they should sleep some distance from where they are fed.

The feed should be composed of a variety and should be as nearly a balanced ration as possible, containing the proper amounts of both fat and bone-making material, and never solely an all-corn ration. Of course, corn is the cheapest feed one can use, in the corn belt, especially if he grows it on his own farm, and if this must be used for the sake of economy, it can be to the amount of about three-fourths of the ration, with the other fourth composed of feeds containing a high percentage of protein. Tankage fed in the proportion of one-tenth to nine-tenths corn, makes practically a balanced ration. Hogs on alfalfa or green feed, need less tankage with the meal, which can be fed either dry or soaked and fed as a slop.

With this ration a feed of the third cutting of alfalfa hay, which is always bright and green, would be an excellent addition, giving both bulk and green feed for the sows.

A mixture of one part shelled corn, one part oats, and two parts finely cut alfalfa hay put through a cutting box, makes a nicely balanced ration, with the addition of 5 per cent tankage, or where skim milk is plentiful, use it instead of tankage, in the proportion of three pounds of skim milk to one of grain.

Another good ration is equal parts of rye and barley ground fine and mixed with 25 per cent white middlings or shorts—on account of price of middlings, although middlings are very good—adding about 5 per cent oil meal or tankage. This makes almost an ideal ration.

A small per cent in weight of a good quality bran added to any of the above makes a valuable addition.

One thing must not be overlooked, and that is plenty of clean, fresh water. If it can be had at will, so much the better. If it cannot, it should at least be given once or twice daily, for the hog needs a drink of water as much as any other animal or human being. I have known pigs to walk directly from a wet feed of nice rich slop to a drinking fountain and take a good drink of water, as though they had been fed on dry feed. I really think that the majority of breeders and farmers overlook this matter of letting the hogs have plenty of water to drink.

Further, the brood sows during the season should, if possible, have some kind of green feed or pasture. Of course in parts of the country where there is heavy snow, something must be fed to take the place of pasture. There is nothing equal to the third or fourth cutting of alfalfa for this purpose. This, if cured without being damaged by rains, is practically as green as it would be in June, and is greatly relished. It can be fed in racks, properly made, or it may be run through a power cutter and chaffed and fed with a portion of the grain ration, as above recommended. A mixture of salt, charcoal, wood ashes and ground limestone or slacked lime is absolutely necessary, and if convenient, add also a portion of ground phosphate rock. This mixture adds much in the way of mineral matter that is so necessary in building up the bone and frame of the unborn litter.

Brood sows should have a dry, warm place to sleep, and but few in number—not over ten or twelve—should run together or sleep in one compartment. This is to avoid their crowding or piling up too closely.

The future of the pig depends much, in fact more than is usually realized, on what the dam receives in feed and care before the birth of the litter. "A litter well born is half raised," and there should be no immediate change in the feeding formula for the sow having just farrowed a litter of pigs, from what she has been having during the period of gestation, only after farrowing the sow should go at least twenty-four hours without feed, with what water she will drink, which in cold weather should be given her with the chill taken off; then, a very light portion of the same feed she has been having. If she has been fed a dry feed, it would be well to use the same proportion in the mixture, only feed it as a slop, with warm water during the winter in a cold climate, and cold water if in the southern states.

This feed should be gradually increased as the litter is able to take all the milk furnished by the mother. Usually at the end of one week, if the litter is an average sized one, the sow can be fed all she will eat up clean.

By the time the pigs are three weeks old they will eat a little on the side from the trough with their mother, and if it is desired to push them to the limit in growth, a small feeding space can be arranged as they may feed from a very low, shallow trough by themselves, unmolested by the mother, giving them the same feed given the mother.

During this period of the early life of the litter the sow and litter should take plenty of exercise for the necessary good of the pigs, for they must exercise considerably during each day, ot they will become fat around the heart and die with what is known as "Thumps," which is nothing more or less than fatty degeneration of the heart, which they will certainly have unless they are exercised daily in some way.

As weaning time approaches, which should not be earlier than ten to twelve weeks, in my opinion, the feed may be lessened for the sow and more given the litter, so that the sow would gradually give less milk and have no trouble when the pigs are taken entirely away.

Some breeders have made it a practice to gradually wean the litter by taking them away for a few hours and then returning them to the mother and following this up for a few days until they are taken away entirely.

Others have practiced taking one or more of the most thrifty pigs in the litter away from the mother first, then after a few days a few others, and finally taking those remaining, believing that pigs so weaned would leave the sow in better condition and less danger of swelling and soreness of the udder.

We never have practiced this, however, during our forty years of breeding pigs. By letting the litter suckle until it is ten to twelve weeks old or a little over, the sow naturally is inclined to wean them herself and if she has been properly fed the last part of this period she will practically give no milk at the end of three months or thereabouts.

When the litter is finally weaned the sow should be given a good fresh pasture of some kind with little else during the next month or two or until time to begin to bring her in condition for another season's breeding.

We pastured something like forty brood sows that had weaned their pigs in May, by turning them directly into a fresh white clover and blue grass pasture where there were plenty of shade. They had no grain or slop for four months, but were given daily from a water fountain all the fresh, clean water they could drink.

They did well on this grass and water diet, but later on were fed two or three ears of corn daily to each sow, besides the grass and water, as I believe for best results they should have a little grain. The sows that were bred for fall litters ran in the pasture with the others and were removed into individual lots about a month before farrowing, where they were given a little corn and some slop with plenty of grass.

To have two litters a year, sows must wean their spring litters by May 1st, so as to be bred for early fall farrow, and the fall litters must be weaned in November or early December so as to be bred for March and April farrow. In the North it is not always practical to have two litters a year and we find it often advisable to breed the sows so that they will farrow one litter the first year and two litters the next year, or, in other words, three litters in two years.—Utah Farmer.

Meteorological Report, Longmont

FOR SEPTEMBER, 1918 AND 1919

TEMPERATURES:	1919	1918
Mean Maximum	79.8°	70.20°
Mean Minimum	45.3°	41.60°
Monthly Mean	62.5°	55.90°
Departure from Normal	+2.3°	—4.11°
Maximum	90.0° on 5th, 10th	90.00° on 15th
Minimum	31.0° on 22nd	32.00° on 30th

PRECIPITATION IN INCHES:		
To Date	8.10	16.83
For Month	1.48	1.71
Greatest in 24 hours	0.55 on 2nd	0.48 on 15th
Departure from normal for mo.	+0.32	+0.57
Departure from normal since Jan.1 —4.61		

NUMBER OF DAYS:		
Clear	19	11
Partly Cloudy	7	13
Cloudy	4	6

The world demands good dairy cattle. Raise them. The world demands milk, butter and cheese. Produce them. On account of the war, there is a scarcity of good, high producing dairy cattle. Get in on the ground floor, now, by breeding your cows to a good bull and reap the benefits in a short time.

Is Permanent Farm Labor Possible?

Chas. I. Bray

Associate Professor of Animal Husbandry. Colorado
Agricultural College

FEW questions have claimed so much attention in the public press since the close of the war as the problem of organized labor. Few people hear of the farm labor problem, yet this situation is quite as acute, if not as spectacular.

The old-time hired man is gone. He used to hire out for eight or nine months of the year, and stay at home the other three or four months. He got $15 to $25 a month with board; $30 was big wages. He probably did not get this till after the wheat was sold in the fall.

His job ended when snow fell, he hired out again when the ground was dry enough to work in the spring. He got up before daylight to milk seven or eight cows, give the hogs their slop, and get his team fed, curried and harnessed. He plowed, harrowed, or pitched hay till it was too dark to see, then got the lantern and did chores again. He slept in the attic over the kitchen. He took his weekly shave during noon hour Saturday and went to town Saturday night. Maybe he hung around the barber shop or the general store till they closed up. In a wet town, if so inclined, he stayed around the saloon till it closed up.

If he was lucky in choosing a place to work he did double chores every other Sunday and got off free the alternate Sunday. If he was unlucky, he did double chores every Sunday and the boss got off. He did not remain a farm hand many years. Farms were not hard to get, he soon got married and bought or rented for himself, or went West to carve out his own fortune.

How Can We Fill His Place?

There have been many methods suggested for filling his place, or keeping those who now try to fill it, but the question is more acute than ever. We said that permanent employment all the year round would answer the question. We have tried foreign labor, married labor, higher wages and then more wages, bonuses, better homes, better farm machinery, the ten-hour day, improved bunk houses and so on. All these things help, but still we seem to be as far as ever from a solution of the farm labor problem.

I interviewed a prominent farmer and sheep feeder of Fort Collins who runs three farms of his own, and has a large interest in the Buckeye Ranch between Fort Collins and Cheyenne. Charles R. Evans is a business farmer of the highest type, a graduate of the Colorado Agricultural College, and a leader in many community interests outside of his business life. I asked what his experience had been in keeping permanent help. He laughed dryly. "That is a mighty big proposition to try to solve," he said. "I keep married men as much as possible, give them a good house and garden and a cow or two and pay them as good wages as I can—usually $85 per month. Single men are unreliable. They get huffy too easily and want to quit, or someone down the road offers them $5 a month more, and they leave. Some of them want to quit anyway as soon as they get a little money ahead."

"Do you find it a good plan to offer a share in profits as part of a man's pay?" I asked.

"I have done that with some of my foremen. I believe it makes

them more careful and more attentive to details. One man I had made enough to double his wages. They don't always care to hire on this basis, but prefer a guaranteed wage as my present foreman does."

I asked if he knew many others who had tried out this system of managing farmers. "No, there are hardly any who try to manage more than one farm. They usually prefer to rent their other farms on shares. It is less trouble and usually more satisfactory. When a poor year comes like this one it is not easy to pay high wages when they have to be dug out of the ground, so to speak. If a cash rent system is followed, and a bad year comes, the tenant loses everything. By the share rent system owner and tenant divide, whether the year is good or bad, and it is not so hard in the latter case on either party. I would not fool with more than one farm if it were not for the greater opportunities they give me for feeding cattle and sheep during the winter."

I had an opportunity, not long ago, to visit the Pine Valley Dairy and Farms Company along with some 60 or 70 farmers and their families. This farm is out among the foothills near Colorado Springs in the shadow of Pike's Peak. It is the home of one of the greatest herds of milking Shorthorns west of the Mississippi, and one of the most beautiful places imaginable. The long, brown barns built on high stone foundations seem almost a part of the eternal hills themselves. One young cow, Victoria 228067, has just made her required year's production of butterfat in five months, and is in full swing after the yearly record for the breed. I had asked Bill Lauck, the county agent, about this farm labor situation during the trip. As we drove into the yards I noticed a nice brown bungalow to our left, not so large as the manager's house, of course, but good enough to be found on any residence street in town. "Here is one of the farm hand's houses," said Lauck.

It Can Be Done

The manager, Seeley G. Rose, was a very busy man about that time, but I got to meet him later on. I asked him about keeping permanent employes. "That is a question we have not entirely settled yet," he said, "but I hope to have it solved satisfactorily when our plans are in full operation. If we can give a man a nice home to live in, pay him a good living wage, and then give him some chance to share in the profits made on the farm, he will be more likely to stay with us. A contented worker is a better worker, and will have a greater interest in saving labor, time and equipment if he has a personal reason for helping the farm to pay dividends."

I do not know whether Mr. Rose will succeed in solving the farm labor problem. I do know where I would go to try for a job if I were to turn farm hand again.

I asked another man, R. W. Clark, now in the extension service, how to keep permanent help. Professor Clark has been farm hand, college professor, farmer and creamery owner. He said: "Get a married man; give him a house, garden, cow and chickens, and make things congenial to all concerned. Give his boys work to do on the farm. The bigger the family he has the better. He won't want to move so often."

I asked if he knew many cases where this method has succeeded in keeping men permanently. He thought a minute. "I don't believe I do."

"Well, what is the solution of the labor problem if a good home and perquisites won't help?"

"I don't know. You see, as Herbert Quick says, if a man is the steady, intelligent, hard-working man you want, he is going into business for himself. The man of the shifting, floating type finds it hard to overlook the high wages paid elsewhere for day labor. He can go into town in 30 minutes and get work with the city street department any time he wants to at $6 a day. Eastern Colorado wheat farmers are offering $8 and $9 a day for harvest hands. Kansas farmers are offering $15. Five hundred dollars or more in two months looks pretty big to the average farm laborer."

The Hired Man's Side

I wanted to hear what the other side had to say, so put the question up to a hired manager. We wanted to know what was most needed to keep good men on the farm. "Well," he said, I don't know that I could tell you. There are so many things that enter into the question. It may be the wages or the living accommodations or the personal relationship between the men and the boss, or the nature of the work." I asked about the bonus system based on profits. "To my mind the bonus is usually a Jonah. Generally the wage that goes with it is smaller and the chances are barely even that a man will gain much. If one could get a good living wage, however, and then a working interest in the concern, I believe he would work better."

He was probably right in saying that no set rule will solve the question immediately and in a satisfactory manner. It is not a matter that can be adjusted entirely between the farm owner and the farm hand. Wages are always going to seem high to the man who pays them and low to the man who gets them. The farmer is going to pay what he thinks he can afford in order to. get the help he needs. The farm hand will. accept what he thinks he can afford to work for; if he does not get it on the farm he will go elsewhere, other things being equal. If the employer cannot, as Mr. Evans phrases it, "dig enough out of the ground," to compete with the "protected" manufacturer in the matter of paying wages, farm production will decline. When values reach a point where farming is again profitable, farm wages will increase correspondingly.

Anything that will make a man feel that he has a home, instead of just a place to stay in, will help to keep him on the farm. Good dwelling houses, perquisites, such as a cow, chickens and garden, bonuses, fair treatment and regular hours, all count towards this end.

If the neighborhood is a desirable one to live in, has good roads, with schools and churches within reasonable distance, a man with a growing family will be more likely to stay and it will be easier to hold the right class of labor. Of course the wage paid, compared with what a man can get elsewhere, will prove to be the important and deciding factor.

The farm workman should be able to hold up his head among men along with the worker in the bank, store or factory; his wife should have as good a home to live in, and his children as good a chance for an education. To do this, means that we who are interested in agriculture, and who specialize in agricultural problems must help the farmer to produce a maximum of food with a minimum of human effort, to produce what will be most profitable for him, and with the least waste, and to get his product to the ultimate consumer by the most advantageous and direct route. In the end, the best working rule for us all is the one laid down some two thousand years ago, to "Treat the other fellow as you would like to be treated in his place."—The Dairy Farmer.

Value of a Pure-Bred Sire

Wilber J. Frazer
University of Illinois

A FEW poor cows may do little permanent harm to the dairy herd, but a poor sire will do untold damage. Frequently dairymen hold the penny so close to the eye it is impossible to see the dollar a little farther off, and this is just what a man is doing who has a good dairy herd of grade cows and thinks he is economizing by buying a poor or even common sire.

If the good pure-bred sire improves the milking capacity of his daughters only one and one-half pounds of milk at a milking, above the production of their dams, this would mean an increase of 900 pounds of milk for the ten months or 300 days an ordinary cow should give milk. The daughter would also be a much more persistent milker, that is, would give milk for a longer time in the year, and she would regain her flow of milk better after an unavoidable shortage of feed, as in a summer drouth. These daughters may certainly be credited with 1,000 pounds more milk per year than their dams produced. At the low estimate of $1 per 100 pounds this extra amount of milk would be worth $10 per year. The average cow is a good producer for at least six years, or until she is eight years old. It will on the average be four years after purchasing the sire before his first daughters will have brought in the first extra $10. Eight dollars and twenty-three cents kept at compound interest for these four years at 5 per cent will equal $10, so the daughter's improvement or increase of income the first year is worth $8.23 at the time her sire is purchased. The cash value of the daughter's improvement (inherited from the sire) figured in the same way for each of the last six years she gives milk is shown in the following table:

Improvement first year	$8.23
Improvement second year	7.83
Improvement third year	7.46
Improvement fourth year	7.11
Improvement fifth year	6.77
Improvement sixth year	6.45
Improvement for six years	$43.85

The total increased income of a cow over her dam by having a good sire is therefore $43.85.

In the ordinary dairy herd of thirty-five to forty cows an average of seventeen heifers per year should be obtained, and twelve of these should be worth raising, making it easily possible for a bull to earn twelve times $43.85, or $526 per year. This would amount to $1,578 in the three years that a bull is ordinarily kept in service.

Cost of providing every heifer one good parent:

	Pure-bred	Scrub
Cost of sire	$150.00	$ 30.00
Interest, three years, 5 per cent	22.50	4.50
Cost of keeping, three years	100.00	100.00
Risk, three years	50.00	10.00
Total expense, three years	$322.50	$144.50
Value at end of three years	100.00	30.00
	$222.50	$144.50
	114.50	
Extra cost good sire, three years	$108.00	
Extra cost good sire, one year	36.00	
Extra cost good sire, one daughter	3.00	

Considering the male calves as worth no more than if sired by a scrub, it would then cost $36 to provide one good pure-bred parent for the twelve heifer calves which are raised each year, or $3 per heifer. Where else can such an investment be found? Three dollars expended brings in an average return of over $7 per year for six years, or $43.85 in all. This makes a clear addition of $43.85 to the income of each daughter, or a net profit of $40.85 and of $1,470 for thirty-six daughters in the three years. Here is nearly 1,000 per cent profit on the investment. The original cost of the good sire looks very small beside the $1,470. It really pays, as nothing else on the farm pays, to put $150 into the right kind of a dairy sire that will return practically ten times $150 within three years.

An examination of details will show these estimates to be conservative. There is plenty of margin left for failures and unfavorable conditions. One thousand pounds of milk per year is a conservative estimate of the improvement of the daughter's production to credit to a good sire, but the details of figuring it may be varied to suit conditions in different herds and different localities. One hundred and forty dollars is certainly a liberal allowance for the purchase of a pure-bred sire, and results here named are based upon having a first-class animal at the head of a herd. A herd of only thirty-five or forty cows is taken for illustration, while a vigorous sire properly fed and exercised is sufficient for a herd of forty-five to fifty cows, provided he is not allowed to run with them. There is another distinct improvement of the good sire's daughter besides her milk production; it is the improvement of her blood or breeding, as the result of which her daughters will be better milk producers. This blood improvement of all the daughters accumulated through a series of years means a remarkable increase in the efficiency of the herd.

It is the common experience of all dairymen who have used a really good improved dairy sire that the investment has made them royal returns. The $150 cost price looks "too big" only to the narrow vision that cannot see the natural improvement of the herd certain to follow. Many a dairyman might have to say that he cannot afford to pay a big price for a fine cow, but the same argument does not apply at all to the purchase of an improved bull, because the sire's influence spreads so much farther and faster than that of the cow.

If the heifer calves are to be raised for dairy cows there is absolutely no business reason on earth for keeping a scrub bull. The dairymen who think there is, pay a heavy price annually for maintaining that tradition. The scrub bull is the most expensive and extravagant piece of cattle flesh on the farm. He does not stop at being merely worthless, but will lose the farmer the price of two or three good bulls every year he is kept. The dairyman could not afford to keep a scrub bull if the animal were given to him, if he were paid for boarding the beast and given a premium of $100 per year for using him. The presence of the scrub in so many Illinois herds—many times without a single qualification except that he is a male—is an offence and disgrace to the dairy business and a plain advertisement of the dairyman's thoughtless bid for failure. The only thing on earth the scrub sire is good for is sausage, and it is high time that this plain and simple truth were given practical acceptance on every dairy farm.

By all means get a good sire if you have to sell two or three cows to do it. The improved sire is, without question, the most economical investment in any dairy herd.

In 1915 the management of the Napa Hospital Holstein herd, Napa, California, undertook to improve its production by the purchase of pure-bred sires. Today we find the eighty-four cows constituting the herd producing a daily average of 3,000 pounds of milk against 2,400 pounds from the 140 cows that constituted the herd in 1912, an average daily increase per cow of 18.57 pounds of milk.

In 1889 a Wisconsin farmer decided that his ordinary "beef and milk" cows were not paying, and he purchased the best pure-bred sire he could get, mating him with his heifers, which were of mixed breeding. By keeping careful records he found that the cows resulting from the first cross averaged a production of 212 pounds of butterfat the first year. By continuing the use of a pure-bred sire, he has increased this amount to an average of 268 pounds at the end of the ninth year.

Bulletin No. 165, issued by the Iowa State College of Agriculture, gives the partial report of an investigation which had been under way for eight years when issued in May, 1916, and shows that the average of all records made by first-generation heifers from scrub cows shows an increase of 71 per cent in milk and 42 per cent in butterfat over their dams, at an average age of three years.

Even at a high price, the pure-bred and bred-for-production sire is better than a scrub at any price. A bred-for-production sire is worth more, not because he is purebred, registered, and high-priced, or has a highsounding name, but because he has the ability given him by generations of careful selection and mating, to produce offspring of equal or greater excellence.—The Utah Farmer.

The Long Term Lease

Oscar Cheairs, Iliff, Colo.

Some time ago I was asked the question, as to which I would prefer, short or long-term leases on farm lands.

There is much to be derived with long leases for the tenant as well as the land owner.

I know of a case in this district of a tenant that has been on the same place for eight years, he has kept the place up in first-class shape, raising good crops every year, and this fall he bought one hundred and forty-four acres of part of the place he was farming at two hundred and twenty-five dollars per acre, making a good payment down. He is now building improvements on the place that will cost seven or eight thousand dollars.

In the first place, under irrigated lands, it often takes the renter a year or more to understand the slope of the land, how to take advantage in running water, and often he can establish small ditches that would be of much benefit in irrigating.

As a general thing, if a tenant knows he has a long-time lease, he will take more pride and interest in the place in the way of making minor improvements, better plowing, and the hauling and scattering of manure over the place in the fall and winter months.

He will also take more pains in the repairing of fences and buildings and the keeping down of weeds along fences and ditches, if he knows the place is in his care for several years, he would not be so apt to call on the land owner for every dollar used in making small improvements, and for small repairs. I have often known them to do a good deal of painting, varnishing and papering of the house; also do the repairing of the barn and other outbuildings. Some may think if they get an undesirable renter it would be hard to get rid of him. This can generally be easily overcome by the provision made in most contracts for such purpose.

Increasing Wheat Yield

Dr. R. J. H. DeLoach
(Address at Wichita National Wheat Show)

 THE largest yield of wheat on record in the United States Department of Agriculture is 117 bushels per acre on eighteen acres in the state of Washington. The next largest is 108 bushels on ten acres in the state of Idaho. These high records would have little significance except for the importance of the wheat crop and the gradually increasing surplus wheat lands of the United States; also the increasing popularity and the growing needs of larger and larger crops of wheat and other grains. This great country of ours can never grow maximum crops on all the lands, but plans can be laid for larger yields per acre.

If every acre had done as well in 1919 as our best acres did, we would have more wheat in the United States than the whole world could use. The average yield per acre in this country is only 15.8 bushels, or less than a third of our good yields. The world's total wheat crop is a little over four billion bushels. If we produced in the United States sixty bushels per acre on the 71,000,000 acres planted for the present crop, we would have sufficient wheat to supply the world. But this is not necessary. Other countries can and do raise wheat and perhaps more cheaply than we do.

This leads us to the question of cost to produce. The Minnesota Experiment Station found that it required twelve and a third hours of human labor and thirty hours of horse labor to produce an acre of wheat. These figures were secured in an investigation carried out in 1909-1912, under the direction of Thomas Cooper. Since the cost was based on hours of labor rather than measured in dollars and cents, it holds good today, and to translate it into dollars and cents, one needs only to multiply the hours by the price of labor at a given time.

If the yield was twelve and a third bushels per acre, each bushel would require one hour of human labor and two and a half hours of horse labor. If the yield had been twenty-four and two-thirds bushels, the cost would have been reduced half per bushel except the very small additional time required for harvest and threshing.

The higher yields are always secured at greatly reduced cost per bushel. The future wheat growers of America must keep constantly in mind methods of increasing the yield and thereby reducing the cost to grow. * * *

One way to reduce cost of wheat production is to select seed wheat successfully, and gradually eliminate all poor and low-yielding plants. My own observations prove to me that at least one-third of the plants in any field are failures—that is, they are not paying the expenses of tilling the soil in which they grow. They get the plant food elements from the soil and fail to give the proper return. Science is helping to eliminate this terrible loss by making more of the plants pay—by making more of them yield full, plump heads—and there is sufficient proof that this can be done.

A second principle to observe in the successful growing of wheat and reducing the cost of production is the proper rotation of crops, whereby cover crops can be plowed under one year in two, or two years in three. Each section of the country has its own crop rotation system, but the plan must always provide for a substantial production of organic matter to be returned to the soil. Rotation helps also to keep the land free from disease, and doubly insures higher yields.

A third point for consideration is the proper balancing for plant foods in the soil, so that the plants may draw upon this source for constant diet. In all the European countries and in much of our own country, much study and thought have been given to the application of plant foods to soils for the purpose of lowering the cost of production. Belgium has perhaps been most successful along this line, and just prior to the war reported the highest yield per acre of wheat of any country in the world, also the lowest cost per bushel to produce. With a wonderful system of tillage, Belgium applies more plant food per acre to her wheat lands than any other country on the globe. This we believe worth serious consideration, in view of the fact that the 10,825,000 tons of straw alone harvested in Kansas in 1919 showed a total of 100,000 tons of potash, to say nothing of the phosphorus, nitrogen and lime.

Wheat is essentially human food. Most of the other grain is fed to livestock, but it is not so with wheat. Not more than 20 per cent of our wheat crop is fed to livestock, while more than 90 per cent of the corn is used in that way, about 95 per cent of our oats, 80 per cent of the barley, and the other small grain crops in proportion, are fed to livestock.

Why is this true? Because wheat tastes better; not because it is particularly better for us, but because we like it better for food. The other grains we eat then in the form of meat, while we take wheat as our bread to go with the meat, and it is well to keep in mind that 70 per cent of the grain crops goes into meat before it is ready for human food, while the other 20 per cent is eaten directly.

The greater part of our farm labor, then, is spent in growing and preparing food for our animal friends, and herein lies the secret of success in farming. Only in this way we are able to keep most of the fertility of the land and sell off a minimum. The animal manures contain our large grain crops, slightly reduced, but with much of the soil-building qualities retained and some of them even improved. This is one of the reasons why wheat farmers should also be livestock farmers.—Dr. R. J. H. De Loach, Director Armour Farm Bureau, in address at the Wichita National Wheat Show.

Western Shorthorn Breeders' Association Sends Out Call

To the Members of the Western Shorthorn Association and Breeders of Shorthorn Cattle:

Another harvest is about finished and we regret that many sections have not had the bountiful crops and abundant grass, that they had hoped for. However, the Shorthorns, wherever found, have done well, under the various conditions, and are in greater demand than ever before.

The membership of our association is increasing, although there are some breeders of Shorthorns in the west that are not members. We extend an invitation to them to join our association. Annual dues $5.00 (payable in advance).

Western Stock Show

The American Shorthorn Breeders' Association announces that Mr. W. A. Cochel will have charge of the Shorthorn Sale, at the next Stock Show in Denver. All members having animals that they wish to sell, in the ring, that will not come up to the sale requirements, should bear in mind that our annual sale held in the spring, furnishes an outlet for such cattle. Our first annual sale held in Denver last spring was a marked success, and the Western Shorthorn Association will hold a sale of this kind every spring.

The buildings that have been purchased and additional space acquired by the Western Stock Show, will relieve the former congested condition and afford better accommodations for the pure-bred cattle.

We wish to call your attention to the greatly increased premiums on Shorthorns. (If you have not received a premium list, send for one.) For the western breeder a futurity class has been put on, and we hope to see a good showing there.

The American Association with its able directors and field representatives are behind us and it is now up to us to show that such recognition is deserved.

In the classification of the specials in the premium list of the Denver Stock Show for western cattle, there were two prizes offered for the same prize winner. We have divided those prizes as follows:

$50 Suit of Clothes by Powers-Behen Co.—Senior Bull Calf.
$50 Merchandise by Denver Dry Goods Co.—Junior Bull Calf.
$25 Loving Cup by Daniels & Fisher—Senior Heifer Calf.
$25 Merchandise by A. T. Lewis & Son—Junior Heifer Calf.
$10 Spot Light by Denver Auto Goods Co.—Calf Herd.

The Future Outlook

The interest in pure-bred Shorthorns is growing at an unexpected rate, in the west, northwest and south, and we should take advantage of the present situation, and not allow this business to go further east. The Pacific Coast, also Idaho, Montana and Utah are forging ahead, and increasing appropriations at fairs, and are building additional accommodations when needed, and your attention is called to these evidences of activity, to show what is being done in Shorthorn circles in more or less virgin territory.

Besides the many Shorthorns sold in less than carload lots, we wish to say that one party in California, and another in Iowa each purchased a carload of females from Colorado breeders.

The Milking Shorthorn

We wish to say a few words about the Milking Shorthorns, and quote Mr. C. H. Hinman, who has one of the largest herds of milking Shorthorns in Colorado. He says that a Holstein holds the highest butter-fat record in Colorado, a Guernsey second and a Jersey third, but these Association reports show that the cows that produce the greatest net profits, under farm conditions, and with farm feed and farm care, are the Milking Shorthorns. They lead every breed in net profit, the leading herd was Shorthorn, the leading cow a Shorthorn. Mr. Hinman says that they sold two barren females in June on the Denver market, for beef, one a cow weighing 1,500, the other a heifer, weighing 1,310, both bringing 12 cents a pound, and topping the cow market for over a month.

The greater percent of Shorthorns are good milkers, and this breed can be termed THE FARMER'S COW.

Write our Secretary, when you wish any information, when you wish to buy anything and when you have something to sell, and also when you have any Shorthorn news. It will assist him to get out a more newsy letter. Let us all pull together, for what we think is the best breed on earth.

In conclusion, we wish to say that we hope to see many new exhibitors at the coming Stock Show in Denver this coming winter. To those that have never shown cattle, we will say that it is one of the most instructive, pleasing and profitable undertakings imaginable to anyone who loves livestock. With proper selection, feed and fitting, you will be successful.

Don't forget that at both the Denver sale during the Stock Show and at the annual spring sale, there will be a sifting committee to bar all animals of undue merit. The pure-bred breeders of any breed cannot afford to sell any animals that are not a credit to the breed.

News Notes

New herd bulls have been purchased to assist the present herd sires in the following Colorado herds, and we are looking for splendid reports from them in the future.

The Allen Cattle Company of Colorado Springs have purchased an exceedingly choice calf from the Owen Kane herd, named Meteor.

The Cornforth Livestock Company, of Elbert, purchased a roan calf of great merit of Tomson Bros., Kansas, named Royal Whitehall.

Messrs. I. J. Noe, of Greenland, secured a roan yearling from the herd of I. M. Forbes & Son.

Mr. Charles Plageman of Eckley, Colo., and Mr. F. A. Galbreath of Yuma, Colo., are two new members of our Association.

Mr. Goss of Fort Collins and Mr. Allen Carnahan of Elbert, Colo., made good selections of breeding animals at the recent sale of I. E. Crews, Haigler, Neb.

Jay & Allen of Boulder purchased three head of cows at the Crews sale. Mr. Jay was formerly of the Jay farms in Iowa.

Harmon Bros., Boulder, and D. Warnock & Sons, Loveland, were also bidders at the Crews dispersion sale.

The demand for females has never been better. If you have any to sell, notify the secretary.

We regret that the members of the Western Shorthorn Breeders' Association cannot well get together more than once a year. However, we hope to see all at the Denver Show and at our next annual meeting.

Yours for more and better Shorthorns,

A. G. CORNFORTH, President.
W. L. WARNOCK, Secretary.

The livestock industry will not have met its full duty until it has placed on every farm in the land pure-bred sires of all farm live stock.—Governor Lowden.

THROUGH the LEAVES

DECEMBER, 1919

THE GREAT WESTERN SUGAR COMPANY

THROUGH
THE
LEAVES

DECEMBER, 1919

Published Monthly by
The Great Western Sugar Company

Longmont Call Press
1919

TABLE OF CONTENTS

Notes on Labor

H. Mendelson

Labor

Up to the beginning of November our farmers delivered in the Colorado district

> In 1919 about 1,230,000 tons
> In 1918 about   980,000 tons
> In 1917 about 1,060,000 tons

In the Scottsbluff district the deliveries at the same dates were:

> In 1919 about 395,000 tons
> In 1918 about 305,000 tons
> In 1917 about 233,000 tons

In both previous years, 1918 and 1917, the fall seasons were as unusually favorable as the year just passed. This is at least one industry where everybody apparently worked at top speed, and far speedier than ever before.

Nevertheless, complaints come from some farmers about the deficiency of the labor. From the above figures it is perfectly obvious that such complaints are not justified except in rather few cases.

There were about 100,000 acres more planted in 1919 than in 1918, and about 58,000 acres more harvested. In the spring it was therefore necessary to provide about 10,000 more beet workers than the year before. The burden of this job was almost entirely borne by the Great Western Sugar Company. It is clear that this extraordinary demand could be only supplied by taking a large number of inexperienced workers where they could be found—namely, among Mexicans.

All other available centers containing German-Russians or Belgians, Philippinos and Japanese, had been utilized to the fullest extent, although only a limited number were available.

That, under the circumstances, some not entirely satisfactory laborers were obtained is really not surprising. Those whose memory goes back ten to fifteen years will remember that at that time complaints about the Russians were quite frequent. There are quite a number of good Mexicans and with a little consideration and patience some of the inferior ones can be improved and others weeded out entirely.

The Sugar Company is going to devote well-organized efforts to this end and, also, is going to take steps to open up new reservoirs of labor similar to the Russians.

Both the company and the farmer are equally interested in getting the best kind of labor. Perhaps both parties have to learn a little more how to go about it to obtain the best possible labor.

585

It is true that some Mexicans broke their contracts. It is true that some Mexicans defaulted in their grocery bills. However, we tried to obtain a record of all of these defaulted grocery bills. The total amounted to less than $2,000, in our districts. This is probably no more in proportion to the total than with other classes of the population. In one grocery bill we noted that the Mexican had been charged 75 cents for a dozen bananas, while the same article was charged to other customers at 35 cents. Some grocery bills contained an undue amount of raisins.

Instances have come to our notice where the farmer had signed a contract in the spring setting a rate of $13.00 per acre for topping. When harvesting time came and the farmer saw that his crop was not going to be up to his expectations, he refused to let the help start work unless they agreed to do the topping for much less than the contract rate.

There were some cases where the farmer had hurried the help to top speed. The harvest was finished before November 1st, but the help had to wait until after November 15th for their pay. Many farmers under these circumstances were able to satisfy the help in some way or another. One case was reported where the farmer could not raise the money but allowed the Mexicans to pick of the turkey flock. The Mexican is still smiling. In fact, there is no doubt that an overwhelming number of farmers tried to do the square thing by the help and in many instances do more than the contract calls for.

However, the few instances where this is not the case create an undesirable disturbance and furnish ammunition for agitation out of proportion to the significance of such isolated cases.

As we must get busy very soon to procure help for next season, it would be well for the farmer to discuss what the pay for the help should be for next year.

We have made arrangements to have all of our fieldmen instructed in Spanish during the next few months so they can better deal directly with the Mexicans. We hope that many troubles due to lack of understanding can be minimized.

"Dairy bulls are judged by their ability to increase the production of their daughters over the dams. Scrub bulls can only decrease production—thus lowering the efficiency of the herd. Their harm is not ended in one generation, but continues indefinitely. With beef at exceptionally high prices sell the scrub bull, for his meat value far outweighs his dairy worth. A common practice for the farmer with a few cows is to breed to the nearest bull regardless of breed, breeding, or conformation and as a result dairy herd improvement is slow."

586

From Factory and Field

Fort Collins

Both last fall and the one now passing have been stormy at about that period when beet tops are ordinarily pastured. The lesson we should learn is the futility of attempting the economical use of these tops by this method. It is a wasteful method at best, unless there is much other pasture available, for the reason that it is too unbalanced a ration for feeding stock, to say nothing of the loss of forage by tramping and the possibility of injury to the soil by packing or puddling. No farmer would think of cutting hay or any other crop and then turning in the stock to clean it up.

Beet tops are recognized as being worth fully as much pound per pound as alfalfa hay, and the present price of forage warrants taking the very best care of them. In riding through the country one can see many examples where a little energy expended would have saved much feed. Much of our beet help top the beets so that the tops are left in a continuous row in the field, and when the stock is turned upon them, as they have been recently, a large percent of the feed is lost that might have been saved by raking into piles. A man and team may easily, by means of a rake, pile twenty acres per day, and even if the farmer does intend to pasture the crop, this work should be done at the proper time. It relieves much of the loss occasioned by tramping should the weather remain clear, and if it snows it makes the crop available to the stock when otherwise it would be lost.

If a farmer is not in a position to silo the tops, certainly it would seem to be a paying proposition to rake them into piles, rounding them up with a fork afterwards. Feeding them to stock by hauling, as all other roughage is fed, and thus be on the safe side by having a valuable feed in such a position that it may be utilized at any time and with the least loss should be more generally practiced where the tops are not siloed. If it will pay to cut and stack or silo and rehandle other roughage, it certainly would pay us to at least handle once, a forage which is recognized as being as valuable as are the tops from the beet crop.—H. H. Griffin.

Billings

The Billings territory has experienced one of the worst harvest seasons this year we have ever had. October 22nd, when we had our first serious snow storm, about 30 per cent of our crop was still in the ground. For four weeks no beets whatever were dug and only by the best of good luck we are now able to dig at least a part of that portion that was left in the ground. Fields and roads are in bad condition, which makes the operation much more expensive than it would have been under ordinary conditions.

Quite a number of our growers who started as soon as they got orders and kept going, had all, or nearly all, their crop in when the bad weather came. Those who had finished their harvest could enjoy a foot or two of snow, for we surely needed the moisture it contained. We are assured of plenty of snow for next year's water supply, and there is plenty of moisture in the ground to start fall and spring crops in good shape.

Judging from this year's experience, we should start our harvest at least ten days earlier than we have been in the habit of doing. November 5th ought to be the latest anybody should figure to have beets left in the ground. In our northern climate, previous experience has shown it is not safe to figure on a later date. As a rule, most of our beet seed here is planted in May. A large proportion of this ought to be planted in April. By planting as early as possible, we get the full benefit of the whole growing season and our beets will ripen earlier, which gives us a chance to start our harvest earlier and the grower eliminates all chances of having part of his crop frozen in the ground.—John Maier.

The Corn Crop in Colorado

J. F. Jarrell

WE usually hear Colorado spoken of as being unsuitable for corn growing. The practical farmer in Colorado, however, is realizing more and more that the corn crop is of vital importance in the agriculture of his state.

With the use of suitable varieties of corn successful crops are being grown in every section where general farming is done. In 1914 the acreage planted to corn in Colorado was 462,000. In 1918 the acreage was 527,000. This is an increase of 65,000 acres, which is a considerable item when we consider that these same years were the ones which cover the time of the World War and also the years in which a much larger acreage of wheat was grown because of the high price of that cereal. In fact, the acreage of wheat in Colorado increased 56 per cent during the years of 1914 and 1918, inclusive.

As conditions and prices get back to normal the conservative farmer will be planning on adopting sound farm practices. Experience tells us "that diversified farming is the best practice." The corn crop finds its place in a rotation such as can be practiced in any beet-growing section. No crop fits in better when a field is foul with weeds following grain. Then, too, as the first crop following alfalfa it gives a chance to dispose of the weed crop and surviving alfalfa, whether the corn crop follows grain or alfalfa, the deep tilling which the soil gets is especially beneficial to the following beet crop. Corn ground can be fall plowed and gotten in fine shape for a seed bed for beets the following year. This is especially true if the corn is used for silage, "a stock feed which is becoming more generally used." A few years ago in passing through northern Colorado one seldom saw a silo. Now they are seen on every hand and in some sections two or three can be seen on one farm. Corn is the principal crop used to fill these silos. It is probably the only silage crop which has passed the experimental stage.

Some authorities give an acre of corn a higher feed value than any other feed crop. This is probably true with the exception of sugar beets. The feed value given to an acre of corn which yields 2,000 pounds is 3,000,000 units of energy. We find in this amount of corn approximately 150 pounds of protein.

Hog raising and corn growing go hand in hand. Colorado is becoming a producer of pork and no feed finishes pork better than corn.

The average yield per acre of corn in Colorado is probably 700 to 900 pounds. However, we must remember that this average includes a high acreage of dry land corn.

The average yield of corn from irrigated land would undoubtedly be 2,000 pounds or more. Many 2,300 to 3,300-pound crops are known to have been harvested.

The kind of seed to use for the corn crop is of vital importance. Yellow Dent, Swadley and the early flint varieties are mostly used. Any good early variety which is acclimated to localities usually does well. Minnesota No. 13 has won an enviable reputation in northern Colorado.

The writer does not think it advisable for any farmer to rush into corn growing as an only means of making money. If he is inclined to do that he had better go to Iowa or Illinois where for a specialty crop he will be in a better environment. Find a place for some corn in your farming plans, either to be siloed or fed in the grain, but be sure that when it leaves your farm it is in the form of a finished product. Pork or beef for instance.

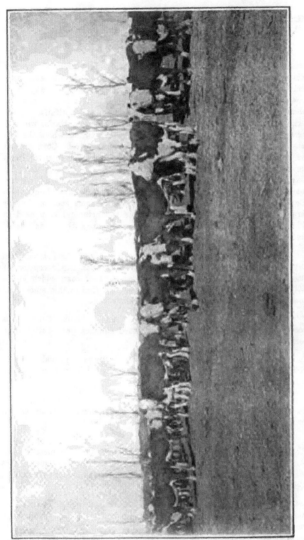

DRY MOLASSES PULP FEEDING EXPERIMENT AT SCOTTSBLUFF (FIG. 1)

Experiments with Different Rations with Dried Pulp as Basis

A. H. Heldt

Agricultural Superintendent, Scottsbluff, Nebraska

We are carrying on at Scottsbluff factory an experiment with one pen of 110 head of two-year-old white face cattle. This pen weighed in at 751 pounds and have been on feed since October 28th. They were started in on the following ration:

2¾ Pounds Dried Pulp.
12 Pounds Beet Tops.
10 Pounds, Hay.

The present ration is as follows:

10 Pounds Dried Pulp.
15 Pounds Beet Tops.
12 Pounds Hay.

In another pen we have 50 head horned cattle (Fig. II), three-year-olds. This pen weighed in at 1,050 pounds. They also have been on feed since October 28th; they were started in on the following ration:

4 Pounds Dried Pulp.
12 Pounds Beet Tops.
12 Pounds Hay.

The present ration is as follows:

16 Pounds Dried Pulp.
15 Pounds Beet Tops.
15 Pounds Hay.

Both pens are showing up well, making good gains. Have had no losses nor sickness. Dried pulp is fed in bunks 16 feet long, with 10 head dehorned and 5 head horned to a bunk.

We intend to run a picture of these two pens every month and keep a record of their rations. Hope when we get through to have some practical information for the small feeder regarding the feeding of his own tops and hay on the farm.

DRY MOLASSES PULP FEEDING EXPERIMENT AT SCOTTSBLUFF (FIG. II)

The One-Year Tenant System

A. C. Maxson

THE one-year tenant system is the cause of many of our poor farms, poor yields, and much of the friction and lack of co-operation between landlord and tenant.

When questioned regarding their reason for renting on the one-year plan most landlords give one or both of two reasons. Either they are afraid that a tenant will stop the sale of the land if he has a long-term lease or else he is afraid he may get a poor tenant and will be unable to get rid of him.

Bearing in mind the fact that the one-year tenant system is a "skimming system," that is to say, that every one-year tenant will attempt to get all he can out of the farm and put as little as possible into it, does it not naturally follow that a farm rented in this way gradually becomes less productive and less valuable.

If this is the case while your one-year rental system makes it easy to sell your farm, it also is very apt to make you sell at a smaller price than you could get, provided a proper system of farming had been maintained and the fertility of your farm increased instead of decreased.

Why does the landlord fear getting a poor tenant on his farm for a term of years? Simply because the poor tenant will leave the farm in poor condition. On the other hand is a good tenant going to rent a farm that has been skimmed by every tenant occupying it for a period of five, six or ten years? Has it never occurred to you, Mr. landlord, that you are renting your land to the poorer tenant simply because you insist on a one-year lease? In final result is there so much difference between the two systems?

Possibly some of the evils of the one-year rental system can be lessened by the landlord planning the rotation system and directing the planting each year. This, however savors too much of autocracy and does not develop the spirit of co-operation between tenant and landlord so essential to good farming and a healthy social atmosphere on the farm.

So far as the writer has been able to observe the most successfully conducted farms are those owned by a landlord who makes a partner of his tenant. This is done by the landlord furnishing certain things, such as a portion of the seed, a portion of the labor or furnishing cattle for feeding and giving the tenant a portion of the profits for doing the feeding. This allows the landlord to assume the part of senior partner, thus acting as the general manager of the farm.

Three things are essential in order to secure the best results from the farm: Rotation, fertilizers and fall plowing. Under the one-year tenant system it is practically impossible to properly incorporate these into the farming system.

Rotation is the most essential of the three. This means the plowing down of alfalfa and the seeding of new. The one-year tenant does not have the opportunity of properly plowing alfalfa since it should be crowned at least the fall before. In case a growth of alfalfa is to be plowed down as a green manure this must be done in late summer to insure killing the plants and properly rotting the green growth. No tenant who is going to leave the farm the first of November will do this preparatory work for another season. Neither can a tenant who is farming some other farm leave his work and do work on the farm he expects to occupy the following year without more or less loss to himself and his landlord. The only remedy is the long-term lease. Rotation also means the planting of fall grain on most farms. This cannot be done economically where tenants have to work on two farms, perhaps several miles apart, because they are going to move at the end of the season.

The one-year-rental system results in a lack of fertilizer. A tenant who is constantly on the move seldom has more than the necessary work animals in the form of livestock. Not knowing whether the next farm he is going to occupy will be equipped for handling stock, he cannot afford to start in the stock business, especially the pure-bred stock business. This robs him of the incentive produced by good stock.

There are only two forms of fertilizer which pay for themselves on most of our northern Colorado farms: Manure and alfalfa as a rotation crop and as a green manure.

Unless the landlord finances the feeding operations little feeding is done and little manure produced. Unless some green alfalfa is plowed down, the greatest benefit from this crop is not realized. The first is only possible through the co-operation of the landlord and the tenant; the last only through the long-term lease.

In order to plant timely, fall plowing is necessary. It is also necessary in carrying out any system of rotation. The one-year lease system does not make it possible to do fall plowing economically for the same reason that alfalfa land cannot be most economically plowed under this system.

We are entering upon a new era: One of co-operation; co-operation of nations, co-operation of interests and co-operation of individuals. No form of industry, whether it be the production of raw materials, manufacture, or the selling of the finished product can long endure where one part of the industry derives the benefits at the expense of some other. Landlords and tenants are producing many raw products. Neither will continue to prosper unless the other does also.

Let us co-operate in renting our lands.

Representative Livestock Sales at Denver Yards

The "Record Stockman" publishes each day a list of "representative sales" showing the number of animals, weight and price. The following tabulation is the total number of head reported in such sales and is intended to show the trend of prices for the period October 20th to November 20th:

FEEDERS

	$9.00 to $9.45	$9.50 to $9.95	$10.00 to $10.45	$10.50 to $10.95	$11.00 to $12.00
800 to 899 lbs.					
Oct. 20th, Nov. 5th	467	290	888	450	389
Nov. 6th, Nov. 20th	429	409	720	363	......
900 to 999 lbs.					
Oct. 20th, Nov. 5th	207	205	729	846	343
Nov. 6th, Nov. 20th	315	783	787	418	64
1,000 to 1,099 lbs.					
Oct. 20th, Nov. 5th	64	41	74	363	570
Nov. 6th, Nov 20th	112	305	348	283	175

The detail figures show that lower prices prevailed during the second period. A few sales of extra fancy stock brought $13.00. The stock offered in all classes centered at about $10.25 with the exception of 900-pound stock in the first period, which averaged $10.60.

The October report shows about the same average prices, except that choice stock sold as high as $14.25.

The report of a year ago records somewhat higher prices, $10.75 for 850-pound stock and $11.00 for 950-pound stock being the average:

LAMBS

	$12.00 to $12.45	$12.50 to $12.95	$13.00 to $13.45	$13.50 to $13.95	$14.00 to $14.45
40 to 49 lbs.					
Oct, 20th, Nov. 5th	3410	1412	2202	....	....
Nov. 6th, Nov. 20th	3624	2283	6007	....	....
50 to 59 lbs.					
Oct. 20th, Nov. 5th	472	3273	4763	6104	....
Nov. 6th, Nov. 20th	2592	3857	9632	11089	....
60 to 69 lbs.					
Oct. 20th, Nov. 5th	868	858	3308	4846	1515
Nov. 6th, Nov. 20th	420	5129	10674	4380	640

Lamb sales centered at $13.50, showing no change from the October report.

The report of a year ago shows an average of at least 75c higher.

Ewes averaged $5.50 to $6.00 compared to $6.00 to $7.00 reported in October. About $7.00 was the average price a year ago.

Denver a Satisfactory Stock Market

J. R. Wood

October was the greatest cattle month in the history of the Denver stockyards, and shipments of sheep surpassed the corresponding month of a year ago by 138,000.

This is positive evidence of satisfactory results both from the standpoint of the shipper and the buyer.

The following tabulation gives some idea of the volume of business transacted:

Cattle Receipts at Denver Yards, Oct., 1919 (head)	119,635
Largest Cattle Month Heretofore, Nov., 1917 (head)	117,933
Conservative Value of October Cattle	$10,000,000
Sheep Receipts at Denver Yards, Oct., 1919 (head)	448,375
Sheep Receipts at Denver Yards, Oct., 1918 (head)	310,594
Conservative Value of October Sheep	$ 3,250,000
Cattle Receipts, Year to Date (head)	688,800
Cattle Receipts, increase over same period 1918	49,000
Sheep Receipts, Year to Date (head)	1,761,450
Sheep Receipts, increase over same period 1918	380,000

November, 1917, held the record for cattle shipments, but this record is now set aside by nearly 2,000.

Going to the ranges to buy "feeders" will soon be a thing of the past, for the growers, knowing a good market price is assured, are shipping direct to Denver where there are many buyers, and buyers are buying in Denver where there is a selection of stock offered.

This practice stimulates better stock raising on the part of grower, for he soon learns that he is penalized heavily for poor stock or culls. The buyer also has an advantage in that he selects his stock, eliminating the culls, in place of buying the "bunch."

Eastern buyers are being attracted to the Denver market, Iowa, Nebraska, Kansas and Missouri often being represented.

Our rare climate, the wealth of late feed on the ranges and better care in breeding are producing animals in the West which are recognized as the best feed lot material in the world. Our usual open winters with lots of sunshine, pure mountain water, clean alfalfa, beet tops, beet pulp and corn, together with care and attention spells p-r-o-f-i-t for the feeders.

November and December will also be great months at the yards, thus completing a record breaker year.

The Newer Knowledge of Nutrition

H. H. Griffin

THE subject of foods is of two-fold interest to the farmer, for he not only employs them for the sustenance of himself and family, as do others in different occupations, but he calls them to his aid in growing animals with which to produce food to sustain the world in great part.

Dr. E. V. McCollum, of the Wisconsin Experiment Station, and now of the School of Hygiene and Public Health of Johns Hopkins University, has recently published a book with the above title, in which he gives the results of careful and elaborate experiments conducted by himself and associates. This book was presented as an interpretation of this literature in a course of lectures at Harvard Medical School. Nearly 3,000 feeding experiments varying in length from six weeks to four years were observed. This book, published by the MacMillan Publishing Company, has attracted wide attention, and I think no doubt puts before the reader or student the best there is in the knowledge of nutrition to the present time.

Not many farmers have the time nor the inclination to secure and read books of this character. With many there is no convenient method by which they may be secured. A summary is often the most serviceable for a busy reader. Owing to the importance of this subject to the farmer, I wish to analyze this book and try to place before him the most essential parts, trusting that it may thus reach many it otherwise would not, and aid in promoting health as well as in getting better financial returns. In making this analysis I have not always employed the exact words of the author, but for the sake of brevity, have at times condensed without changing the meaning, in order not to make the article so long as to tire the reader.

A plant structure or an animal body is an exceedingly complex mixture of chemical substances. Through a century of patient labor by many able men an understanding of the number of characters or small structural units into which the tissues of the animal or plant can be separated, became realized. Living tissues are known to consist of protein, carbo-hydrates (sugars and starches), fats, minerals, salts and water. Pronounced differences were observed in the composition of many substances which serve as food for man and animal. Meats, milk, eggs, peas, beans, etc., are very rich in protein. Great variations are observed in the water content and in the fats and carbohydrates. One of the first efforts in the science of nutrition was the tabulation in classified form of the chemical composition of an exten-

597

sive list of foods, but until the year 1900 the idea that there was any variation in the quality of the proteins from different sources did not become known or recognized. It is not appreciated, as yet, that any variation exists in the other substances, but from the results obtained in actual practice it is reasonable to suppose that variations do exist and Dr. McCollum has also shown in these experiments that there is a wide variation in the value of fats to sustain growth.

Restricted diets have produced for centuries diseases in man. Scurvy, for instance, is one that appears in our western hemisphere; Pellagra, a scourge among the poorest peasants of Europe; Beri-Beri among the poorest classes of the Orient; a condition of the eyes known as Xeropthalmia, is of dietary origin.

In 1907 Dr. McCollum began the study of nutrition problems at the Wisconsin Experiment Station. An experiment was begun with calves weighing 350 pounds, the object of which was to determine whether rations, so made up as to be alike in so far as can be determined by chemical analysis, but derived from a single plant, would be of the same value for growth and maintenance of vigor in cattle. The ration for one group was derived solely from all parts of the wheat plant; • for the second group the ration was solely from all parts of the corn plant; the third group solely from all parts of the oat plant; and the fourth from equal parts of the wheat, corn and oats. The animals were restricted to this diet except being given all the salt they cared to eat. All groups ate practically the same amount of food and digestion tests showed there was no difference in digestibility. Not for a year or more was there any distinct difference in the appearance, after which there came a marked contrast. The corn-fed animals were in an excellent state of nutrition; the wheat fed were rough-coated, gaunt and small of girth, but the weights did not differ much; those fed the oats and those fed the mixture stood intermediate between the two above; that those receiving the mixture would do better was not realized. The results of reproduction are more marked, the young from those fed on corn were full size, strong and vigorous; those from the wheat were the reverse in every respect, they were small and were either born dead or died in a short time; those fed on the other two rations produced young of full size, but about two weeks too soon and they were weak, the most of them dying later. This experiment was carried over the second season with similar results. The milk produced per day was for the corn fed, 24.03 pounds; for the wheat fed, 8.04; oat fed, 19.38 pounds; and for the mixture, 19.82 pounds. For the second period the corn fed produced 28 pounds; wheat, 16.1 pounds; oats, 30.1 pounds; and the mixture, 21.3 pounds.

This experiment confirmed the convictions of Dr. McCollum that the only way in which the problem of nutrition could be solved would be to solve the problems of the successful feeding of the most simplified diet possible. In this way it would be possible to proceed from the imple to the complex diet employed in practical nutrition, ascertaining the faults of the seed alone and the leaf alone.

A sufficient number of comparable experiments have now been conducted with several species of animals to make it appear certain that the chemical requirements of one species are the same as that of another among all the higher animals. The requirements with respect to physical quality of the food vary greatly. The ruminants (cows and sheep) must have bulky food, while the ominvora (man, pig, rat, etc.), cannot, because of the nature of their digestive tract, consume enough of such foods as leaves and coarse vegetation to meet their energy requirements.

In the experiments where efforts were made to nourish rats on a diet composed of purified protein, carbo-hydrates, fats and mineral salts, the results of the other experiments were confirmed, in that animals can live no longer on such mixtures than if allowed to fast. The rations were of such character that chemical analysis could reveal no reason why they should not nourish the animal. In these experiments the interesting observation was made that growth could be secured when the fats in the food were butter fat or egg yolk fats, whereas no growth could be secured when the butter fats were replaced by lard, olive oil, or other vegetable oils. It was definitely established that, contrary to past belief, the fats are not all of the same food value. Certain fats contain some substances which are not dispensible from the diet, whereas other fats do not contain the dietary essential in question.

Hopkins, of Cambridge, England, conducted experiments similar to those above described and found that neither maintenance of weight nor growth could be secured with the ration first above mentioned. But when he added milk to the extent of 4 per cent of the total dry matter of the ration, he found that growth could be produced.

These experiments brought to light two new view points in animal nutrition—first of which is, that the inorganic content of the wheat kernel, although it furnishes all the necessary elements, does not contain enough of certain of these to meet the requirements of the young animal during the growing period. The second is that the wheat kernel is too poor in its content of an unidentified substance which butter fat contains to satisfactorily nourish the animal over a long period of time. By the biological method (animal feeding) it was shown that seeds have the same faults from the dietary standpoint. That they are similar as regards inorganic content and certain fat constituents, but it is necessary to supplement any mixture of seeds with respect to these before good nutrition can be secured. It is not, therefore, possible to secure appreciable growth in young animals fed exclusively upon seed products.

Our farm animals live for years and thrive upon vegetables, but there is a special property in the leaves that make good the deficiency of the seeds. It is a fact that the hog, an omnivorous feeder, if fed for successful pork production must have access to good pasture along with grain, or else be fed milk in some form with the grain.

The leaves of plants can be classed together in a general way as foodstuffs of a general character, just as the seeds resemble each other in feeding properties. It is worthy of emphasis that in our hundreds of trials with diets derived entirely from vegetable sources we have not succeeded in producing the very best results in the nutrition of an omnivorous animal, the rat. Many people who hold that they adhere to a vegetarian diet in reality take in addition milk and eggs or both. When properly planned this is a most highly satisfactory diet for the nutrition of man. It has been said that it is not possible to make a diet from seeds or seed products that will properly nourish the animal during growth, nor will they maintain a full grown animal over a long period.

For many years it has been pointed out that the protein and energy value and its digestibility were assumed to determine the value of a food, but it may be truthfully said that both the ordinary and the most searching method of the chemist fails to throw any great amount of light on the value of food for inducing growth. The fact that the cereal grains are too low in three organic elements to admit of growth, makes it clear that food packages just as they come from the hand of nature are not necessarily so constructed as to promote health.

In a subsequent number we will analyze what Dr. McCollum has determined in regard to foods of animal origin.

CARE OF FARM MACHINERY

When you have finished using any implement or tool, put it away, protect it from the fall and winter storms.

To properly care for the farm machinery means that it must be well selected, kept in good repair and adjustment, oiled thoroughly, cleaned before housing, and it must have all wearing parts well greased when not in use, and painted when necessary, and it must be properly housed.

At least one-half of "good care" consists in keeping the machinery properly repaired, in good adjustment, and thoroughly oiled when in use. To neglect any of the lines of care mentioned, means serious damage and loss to the machine.

Machinery is going to cost more, the cost of raw material shows this, and it is only another reason why we should be more careful with our machinery.

The only reason we keep saying something about this is because we see so much machinery out in the open, exposed to the wind, sun and storms.—Utah Farmer.

National Western Live Stock Show
1920

Arthur Johnson

Editor Denver Daily Record Stockman

ITH close to 40,000 square feet of added show space this season, the Fourteenth Annual National Western Stock Show will offer a greatly widened exhibit scene to its patrons. Not only in space covered, but in attractions, premiums, horse show features and everything else the show that opens on January 17, 1920, will eclipse any festival and exhibition effort of this nature ever seen in the West.

The live stock exhibits, in fact, promise to strain the capacity of the show buildings now provided. Echoes from the Royal at Kansas City and the International at Chicago carry warnings of a genuine rush of pure breeders to pose their animals in a show reserve where East meets West—where the strains from the uplands of the Rockies mix with those from the gentle confines of the Missouri and further eastward.

The property newly acquired by the stock show management opposite the great Stock Show stadium has been equipped with display buildings of the latest design. Here will be found the Hereford and Shorthorn exhibits heretofore shown in the large barns adjoining the great show ring. The transfer of these cattle exhibits will afford better show room for the large consignments of dairy cattle and hogs which are expected. The latter will be shown in the building formerly occupied by the beef cattle.

Larger premiums, a greater number of exhibits, specially dazzling night shows and matinees in the horse ring and an imposing array of industrial exhibits are to be features of the great 1920 Stock Show. The premiums this year will aggregate $40,000, which is $10,000 more than any total amount ever offered before.

The members of the boys and girls' pig and calf clubs of the state are on tiptoe with expectation. Their prize-winning animals are coming to the stock show. There will be entries from twenty or twenty-five counties. The boys and girls are preparing to spring a pageant in the stadium which will be an eye opener to those who have not been following their work.

The Stock Show management has been giving especial attention to the lining up of eastern stables which will participate in the horse show performances. Several strings of ponies, hackneys, shetlands, and Welsh horses from eastern states have registered for participation in the interesting arena events. Over the tan-bark will sprint some of the best horse-flesh known in the United States.

Unique entertainment and band concerts will enliven the horse show as usual.

The poultry exhibits, the hog exhibits, the feeder displays and the various sales of pure-bred animals will take up the time of the visitor in Denver—such time as he does not spend watching the horse events—during the great show week, January 17th to 24th, 1920.

Feed Cost of Raising Heifers

The importance of providing cheap feed for growing heifers and practicing thorough culling is brought out by the results of feeding experiments recently completed by the Dairy Division, United States Department of Agriculture. In these experiments groups of calves were fed from birth to one and two years and a record kept of all feed consumed.

In one experiment, 11 heifer calves were raised to the age of 1 year. The amount of feed consumed by each calf was as follows: Hay, 571.8 pounds; grain, 885.5 pounds; silage, 3,693.1 pounds; milk, 110 pounds; skim milk, 2,414 pounds. Estimating the hay at $30 per ton, grain at $60, and silage at $8 per ton, milk at 4 cents and skim milk at three-quarters cent per pound, the cost of raising each heifer to 1 year of age was $72.42.

Five of the calves from the first experiment were then fed for one more year. During this second year each calf consumed on the average 1,117.8 pounds of hay, 1,221.6 pounds of grain, 8,031 pounds of silage. Using the same figures for computing the cost of the feeds consumed during this second year, the total cost of raising a heifer from birth until 2 years of age was $157.96. These figures emphasize the necessity of providing cheap feed for heifers, such as pasture and silage, and bring out the importance of carefully culling the heifers to avoid raising those which will prove to be inferior cows.
—Reclamation Record.

IF YOU TREAT 'EM ROUGH

You Deserve to be Left in the Lurch When Labor is Scarce. By Roswell Phillips

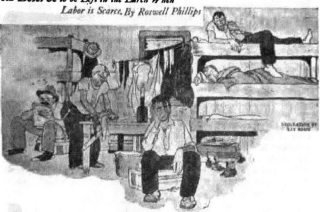

(Reproduced in part from The Country Gentleman)

The stories of tenement life in our large cities pictured by the sob writers of the daily press and by various and sundry social workers never fail to arouse the active sympathy of the average country man or woman.

* * * * * * * * * *

Now I am not going to dispute that there are several million honest farmers nor am I going to enter any plea in defense of sweatshop drivers or owners of mean tenement houses. They deserve all that has been said about them and much more, but if I can I propose to tear the cloak of smug satisfaction from off the shoulders of a lot of country dwellers who are every whit as mean in their treatment of the weak and helpless as any sweatshop driver in the land. All goodness is not centered in the country nor are all the small-souled, grasping, hard taskmasters located in the cities.

* * * * * * * * * *.

Some years ago when I lived in North Dakota I occasionally attended farmers' meetings and one of their topics for discussion was always the hobo labor they were afflicted with at harvest and at threshing time. They complained bitterly of the morals of these men and of their general no-account character. If one listened to these discussions he could not help feeling that these North Dakota farmers had a real cross to bear. I fancy during the last three years they would have been glad to bear it again, but at that time they cried loudly for relief. These men were a menace to the country, they were scarcely human, and generally all their bad qualities were attributed to their city training.

One summer when I lived out there a cry went up from the wheat fields that more men were needed to save the crop. It was in dire danger of being lost for lack of man power; so, like many others, I concluded to do my patriotic duty, which, by the way, came fairly easy on account of the tempting wages offered and the tales of good treatment. The first place I struck was one of the big ranches. I had read of this particular ranch and had seen the pictures of the farm buildings in the magazines and they looked mighty fine. I concluded the people who ran the place must be real folks. When I reached the office I got a job all right enough and was told to go over to the bunk house a mile distant and wait until the boss came in, when I would be told where to begin work.

One look at the bunk house was enough. It was as dirty as a pigpen. The bunks were merely board boxes along the wall with an armful of dirty straw for a mattress. The blankets had never been washed and I doubt very much if the place had ever been swept. Perhaps it was left to the men to do their own housekeeping from year to year, but I am sure the same dirt was there that the first occupant carried in on his boots, and it was then an old building. As I say, one look was enough. I just threw my pack over my back and marched on to a place a couple of miles distant.

The next place was not a bonanza farm. It contained perhaps one section of land, which was a modest-sized farm for that country. I got a job easily enough at satisfactory wages and went in with the men to supper. I remember the flies were very thick, as they usually are at harvest time on the prairies. There were no screens on the windows or doors of the cook house and I noticed a pan of milk in the kitchen black with flies swimming about. When I got to the table the semi-liquid butter was filled with flies, there were flies in the gravy and I found a couple of flies baked in the slice of bread I endeavored to eat. Now I am not particularly squeamish about what I eat, but there were too many flies for me.

When I asked where I should sleep the boss looked at me in a rather surprised way and said: "Suit yourself; you have the whole state to choose from and that ought to be room enough." That was literally the truth; I had the whole state to choose from with only one restriction imposed, and that was not to set fire to any of the buildings. Absolutely no provision had been made to take care of the fifteen or twenty men who were engaged to do the harvesting. Mosquitoes were singing about in clouds and the gnats were plentiful. It was a delightfully democratic and care-free life if one didn't care, but there was not much chance for comfort. I finally rolled up in the haymow, but one of the horses in the stall below must have had either colic or a bad case of insomnia, because it pawed about all night.

From that time to this whenever I hear a North Dakota farmer or any other farmer complaining about hobo labor he gets no sympathy from me.

* * * * * * * * * *

Within the past few months I have seen even worse living conditions for farm help than any I have described. Only a few months ago I visited a ranch in Texas, where every conceivable mechanical device was employed to do the work and apparently every effort was being made to make the plantation efficient—with one exception, and that was the method of taking care of the help. There was one particularly poor old building standing at a crazy angle in the barnyard that I supposed was some sort of storeroom until one of the men who came with me said: "When the boss is not looking just peep into that building." I did and found that it was a bunk house. The bunks were three deep. There was no bedding except some dirty straw and ragged blankets and these were filthy, intolerably filthy. No man with a spark of decency in him would tolerate such a place. There are different degrees of dirt; some is just plain, honest dirt and some is filth. This was filth, and an accumulation of filth that had come down through generations of farm hands. The house had one window and one door. There were no screens, and if the men kept the mosquitoes out, no ventilation. I know of no cellar basement in the cities or back-room tenement that could be worse. This building, moreover, was set in the edge of the cattle yard and during the winter months was surrounded by a lake of mud. No, it was not a negro cabin; it was intended for white laborers.

But speaking of negro cabins, they are not just what might be called strictly sanitary. Of course, I know very well that an ignorant negro family spends very little time in keeping house. They might have things better even with their poor surroundings, and I am not going to criticize the Southern plantation owners because they do not provide their tenants with modern conveniences. Nevertheless, it seems to me the human animals should have an even break with the dumb animals. Even a negro family ought to have as good quarters as the mules or the cattle, but in thousands of places their living

605

quarters are worse. The Brazos bottoms in eastern Texas furnish a very good example of what I mean. There are two men down in that country whom I shall call Boswell and Milnor, each of whom raised a large valuable crop of rice.

Boswell is a hard-headed individual who was born with the idea that he owes no man anything except the wages he agreed to pay. The matter of treatment never occurred to him as a subject worthy even of consideration. He furnished some sort of place to sleep and food of a sort and that was all he felt called upon to furnish. He is the master and people who don't like his style can go—well, they can go elsewhere. I am told the meals he furnishes are noted particularly for their low cost and lack of variety. The houses are tumble-down affairs without screens, often without glass in the windows, dirty and unkept.

Milnor lives perhaps twenty miles away from Boswell and is exactly opposite in character. He is equally careful as a business man; he is noted as an exceptional farmer in his methods of cropping and caring for his land and he has many warm friends. He employs a number of men throughout the year and at harvest time the number is considerably increased. His laborers' cottages, even for the casual laborers, are models of neatness. He employs a cook to prepare the meals, good beds are furnished with clean bedding and the men have access to a shower bath every day.

As long as there were more men than jobs Boswell got along very well. He had men whenever he needed them to take care of his harvest. He proved this theory every year and laughed at his neighbor for being a soft head. But last year he paid the price. Labor was scarce and could choose. Milnor with his neat living quarters had all the men he needed and they were content. They worked well and he got his crop in before the rains came. Boswell stormed and fumed. He cursed his help and the elements, but mere words never did cut any rice no matter how sharp they are. He lost his entire crop of more than 300 acres, worth at least consideration $18,000.

The day has gone by even on the farms when hired men can be treated like beasts of burden. The practice never did pay even in the worst of times, for the best men invariably drifted to the best places and no one ever made much profit out of low-grade labor.

* * * * * * * * * *

The whole thing simmers down to a question of intelligent self-interest. Labor now is in a position to pick and choose. It would be very foolish to pick the worst places and it is not foolish to that extent. It will go where it can get the best wages and the best treatment and in many cases treatment will count ahead of wages.

Dig Deeper But Save the Foundations

E. Davenport

Dean, College of Agriculture, University of Illinois

THE most casual observer must see that the spirit of investigation is abroad in the land—investigation not for the purpose of finding facts upon which to establish national programs of production and distribution, but for the purpose of prosecuting somebody and putting him in jail.

The whole influence is restrictive and tends powerfully to check production and distribution at the very time when both are most needed; and, what is even more unfortunate, it breeds and fosters an ugly temper among the people, filling them everywhere with distrust amounting to a desire for revenge. It seeks the profiteer in order to punish him and is naturally directed against the producer and the dealer, especially against those forms of big business that deal in necessities and that, therefore, have been strengthened rather than weakened by the war.

That there is profiteering no intelligent man will deny, but that the amount of it is sufficient to affect prices notably is exceedingly doubtful.

The danger lies not in legitimate prosecution of real offenders. It lies rather in the overworking of the principle of pitiless publicity and in its application in advance of real findings, taking the form of widespread and sensational press announcements of proposed investigation, seasoned by the recital of a few surface facts that can be easily so pictured by the ignorant reporter as to seem conclusive evidence of guilt. With many good people an indictment is equivalent to a conviction.

We are just now going through a temporary drop in values of many commodities bought by these methods and to be paid for later with a heavy price by an oppressed public that is wearing furs in August, driving automobiles as never before and spending money like water for all sorts of luxuries that human ingenuity can invent.

It is about time this illegitimate handling of facts should be tempered by reason and a decent regard for public welfare.

For example, a reporter seeking a story wanders to the outskirts of the town and comes upon the city dump. Then he discovers—potatoes! Potatoes thrown away. Potatoes deliberately destroyed when people in that very city are starving! He fishes out a few, takes them home for baking and, presto, he has his story! Tomorrow all the city will be moved with righteous indignation that the dealers should be allowed deliberately to destroy stock in order to produce a scarcity and thereby to raise prices.

High Cost of Hoarding

This curb reporter did not know enough about the fundamental facts of the potato business to think that a carload of potatoes had probably been frozen in transit and for that reason condemned to the dump by the inspector and that by searching he had chanced upon some of the few that always escape a freeze. And he did not know enough of business in general to know that the few potatoes he saw, even if deliberately destroyed, would be too insignificant in amount to have the slightest effect upon price; that no dealer could afford to sacrifice enough to accomplish that purpose; that only a great combination of dealers could do such a thing and even then only in time of scarcity, and that the law of supply and demand insures a satisfactory income to the dealer and makes this unnecessary.

Again, whatever the reporter knows or does not know, a paper that deals at all with this brand of dangerous explosive should know that the blame for the potato incident lay, if anywhere, with the management or lack of management which let the potatoes freeze in transit and sacrificed not only a dealer's legitimate profit but at the same time destroyed a valuable food in time of scarcity. The point is not that a few potatoes should have found their way to the dump, but it is the question whether in general our potato crop is being distributed to the best adavntage consistent with reasonable expense. This is what the public really has in mind and we must insist that he who deals publicly with facts of this dangerous character should dig deep enough to discriminate between surface indications and a real "lead"—in other words, between facts and fancies.

It is methods of the kind just recited that have led the public to believe that vast stores of food and other necessities are held back from the consumers for the purpose of raising prices; and there is just enough truth in the matter to give color to the assumption.

But the public in general does not stop to think that it is tremendously expensive as well as dangerous to hold back supplies. Refrigeration is costly. It ties up an immense amount of idle capital. New supplies are always coming in. The public can substitute one food for another or, as in the case of clothes, go without until, under all the adverse conditions, the dealer is driven to the wall and the market breaks. Hoarding in the hope of forcing a rise is not business. Not only that, but the plan can be made to work only in times of great scarcity. All experience shows that both practices react upon the perpetrator much more rapidly and effectively than can the orderly processes of law. Hoarding and profiteering, therefore, may safely

Sunlight Farm, Lovell, Wyoming

180 Acres Beets in One Patch

be left to take care of themselves except in times of scarcity and great public danger, when extreme methods may be necessary.

It is only in times of excitement that the pitiless publicist comes into contact with wholesale stock or has the slightest conception of the vast amounts of ordinary commodities that go to make up the reservoirs, gathered from the producer, and that feed the streams to the consumer. The reporter or other agitator bent upon reforming abuses has seen potatoes, flour, meat and eggs only in the bushel, sack or other retail container. When for the first time in his life he visits a cold storage plant and beholds those rows upon rows of chilled carcasses stretching away into the distance, he says: "Ha! Eureka! Here is the cause of all our troubles. All this is being held back from a hungry world."

When he discovers the "hair" that covers much of the meat and learns that it is really a mold, he assumes that the meat has been held too long and is becoming unfit for use. He feels justified, therefore, in exposing the packer, not only for hoarding and profiteering, but for peddling embalmed beef to an unsuspecting and innocent world.

Eevrybody has heard of ripening meats, but everybody does not know or stop to realize that meat production is an exceedingly irregular business while consumption is fairly uniform throughout the year. Therefore, refrigeration, though expensive, must be employed to steady the supply. Not only that, but the miles of carcasses mean millions of money invested and lying idle. All this and more the reporter does not know, but without stopping to dig deep enough to find out whether the public is really being served or exploited—all of which is some job—he rushes away to take off the lid, that the public may know how it is being mulcted.

What is true of meats is true of eggs, butter and all seasonal products that are also perishable. The basis of the blunder lies in the laymans unfamiliarity with goods in wholesale amounts and in his vast ignorance of the real and necessary processes of business in gathering up some hundreds of perishable products irregularly produced at the ends of the earth and sending them out in a steady stream every day of the year back again to all the people, reassorted and redistributed and still fit for human consumption. Dig deep, therefore, and deep enough to get all the facts, but let no man dig in the wrong place and thereby undermine the foundations of the vast structure of business whereby we live from day to day.

We are "out" after big business and all hunters love a shining mark. Just now we are hard on the trail of the packers, but the case as it is made up outside the courts consists of a heterogeneous collection of suspicions and criticisms whose chief effect, next to inflaming the public mind, is to obscure the issue and in the end to give the packers a clean bill of health whether they deserve it or not. All this is for the very good reason that, bad as the charges and suspicions look upon the surface, nothing is easier for the packers than to give perfectly natural and satisfactory answers and explanations to all the surface charges. In the meantime the real case lies deeper down than the explorers have yet dug. For example: Part I of the Report of the Federal Trade Commission is a voluminous document. It runs to 574 pages and Part III is already off the press. The very bulk of the expose is ominous and of itself tends to create the impression that something is rotten in Denmark.

Confirmatory of this impression is the summary, which is printed first and whose opening statement is as follows:

"The combination among the Big Five is not a casual agreement brought about by indirect and obscure methods, but a definite and positive conspiracy for the purpose of regulating purchases of livestock and controlling the price of meat, the terms of the conspiracy being found in certain documents which are in our possession."

This is very strong language in advance of court findings.

Very much is made of the fact that the packing industry handles many commodities outside of meat and this is followed by a recital of the Instruments of Control and Monopoly, as follows:

"Stockyards,
"Private refrigerator car lines,
"Cold storage plants,
"Branch-house system of wholesale distribution,
"Banks and real estate."

Why Armour Built Cars

All these things being true, the packers are presumably in a position to control prices. Now, the real question is not whether the packers could control prices, but whether they do control them; and this question the accused may easily avoid by a perfectly consistent explanation of the surface charges.

The packers can truthfully point out what most people do not know, but some of us very well understand, that as Phil Armour, the father of the packing industry, developed his business beyond that of the common butcher, he began—fortunately for the world—to dream dreams and have visions of improved service. He began to think of the time when he should supply the distant markets with dressed beef killed at great central abattoirs rather than to continue shipping double the weight in live cattle to be killed in a multitude of local uninspected slaughterhouses, wasting much of the by-products that had already come to have great value.

But this meant refrigerator cars, and Armour exhausted his powers of persuasion to induce the railroads to build them, only to be laughed at as an impractical dreamer.

Accordingly, Armour was obliged to follow the example of Pullman—that is, to build his own cars and borrow money for the purpose, when he greatly needed every dollar for the expansion of his legitimate business of handling cattle.

What happened next? . Armour's cars discharged their cargoes of meat on the Pacific Coast, where were gathered vast quantities of fruit, salmon and other perishables awaiting shipment toward the population centers. Should he load both ways like any good teamster or should he go back empty, leaving the fruit and fish to rot?

It may be said that the fruit men and fisher folk should do what Armour had done and build refrigerator cars of their own. But we must remember that fruits and fish are seasonal products, not constant, and, besides that, if they of the Pacific slope had followed Armour's plan then both lots of cars would have been empty half the time. Then the public demand would have been for consolidation, and rightly so. Now it is for compelling the railroads to do what they at first declined, namely, to build and operate sufficient cars to serve this need.

In this connection we can raise the question whether the railroads have ever succeeded in meeting the full demand for cars for shipping non-perishables and whether we shall be better off in trusting our meats, fruits and other perishables to them, when they have confessed in public hearings that the average daily journey of a loaded freight car is but ten miles, or half what an ox team can do, and that in times of greatest shortage thousands of cars stand idle on sidings in yards and fields until it is an open question whether the freight car is mainly a wagon or a warehouse.

When the Commission is discussing the advantage of the Big Five over the small dealer in meats, it points out that "the big packers' cars have been carefully handled, promptly returned and used only for the shipment of the packers' own commodities," while "the small packers, on the other hand, have been subject to extreme delays in securing the return of their cars. Six months for a trip from St. Louis to New York and return was not at all uncommon." It is further recited that these cars were sometimes used by the railroad for the "shipment of onions."

And yet the Commission proposes in the interest of weakening a monopoly to turn over the handling of the refrigerator-car service to the railroads! The reader wonders whether the Commission has not forgotten that the principal purpose to be accomplished is to serve the world with meat and whether it has not become more interested in busting the trusts or, at least, in finding ways of tying the hands of business than in seeing to it that the public is efficiently and properly served.

Convicted Out of Court

We are also convicting the packers out of court in the public mind because they deal in so many commodities besides meat, having little knowledge of the variety and magnitude of the by-products or their value to the world and no conception of the complicated inter-relations that are involved in their manufacture, transportation and disposal.

Here again we are not to assume that complicated big business means exploitation, but we are to inquire whether the public is being served or whether it is being exploited.

Of a piece with this is the natural charge that because the packing companies all wire telegraphic instructions to their buyers in the various markets, therefore they are attempting to control prices, and when one company frequently buys one-half of a shipment at practically the same price that another company pays for the other half, therefore the two have conspired to fix prices and have succeeded.

The packers may combine, for all we know, to control prices, but this is no proof of it, for if the buyers are up to their business—and there are no keener buyers in all the world—and if the seller knows the value of his stuff, the two halves of the same lot will naturally go for approximately the same money nine times out of ten, unless the market breaks between the two deals.

An even more foolish accusation and one so easily met by the packers as to be beyond the pale of reasonable consideration is the remarkable discovery that each of the packers maintains almost a constant percentage of the total business and, more remarkable still, this fact is advanced to a credulous world not only as evidence but as proof that they have conspired to divide the business between them.

Is Public Served or Exploited?

Without a doubt a statistical study of the great department stores of any city would show that each has secured and is holding a fairly definite percentage of the total trade in its line. If it should lose much it would be speedily forced out of business, and if it should gain perceptibly it would soon swallow all competitors.

These things are dangerously trivial because they are the things that excite, while from the nature of the case they inhere in big business.

The only real question is: Are the packers serving the public or exploiting it? This question they should be forced to face whenever the public has cause for suspicion, because when private business reaches vast magnitudes it becomes a public service about which the public has a right to be informed. But it is a foolish public that proceeds first of all to smash a big and complicated and useful machine built by long years of hard work and great expense just to see how it is made and in order to get scrap that can be melted down for material with which to rebuild a new machine on a new model. Wherefore, dig deep enough to get at the real facts, but not deep enough to wreck valuable machinery.

The packer claims that he is a public benefactor to the consumer because he has succeeded in sending good meat almost to the limits of the civilized world. He is right, but that does not prove that the service is perfect. He claims the gratitude of the producer because he has made a market for the farmer every day of the year and that to do this he has been compelled to invest millions of money in storage, in refrigeration and in idle stock.

Here he is right again, but beneath these two great facts lies a whole underworld about which the public knows next to nothing and in which even the packer himself may be ignorant of all that is going on, because abuses may creep into any man's business without his knowledge, much less his consent. Wherefore, let the packer, the farmer and the consumer together take stock of the situation and go into the business of actual exploration.

For instance, the farmer points out that while he does have some kind of a market every day of the year, there are a good many days when, if he should avail himself of the privilege, it would spell loss and disaster, whereas a day or two either way would provide reasonable markets. Inasmuch as no man can tell the farmer which days to avoid and inasmuch as his beef is worth as much to the consumer one day of the week as another and inasmuch as the supply of meat must be definitely planned for months, if not years, ahead, the meat business is now too hazardous for the best interests of either the producer or the consumer. The farmer is right in believing that better methods of marketing can be worked out.

Now the farmer is not going to be benefited in this matter by smashing the machinery. What he wants is that the processes that have been built up shall be further refined and perfected—specifically, that we shall approach as near as possible to the contract system of producing meats in which the element of hazard shall be lessened to both the producer and the dealer and in which the consumer shall have the benefit of an average lower price. To accomplish this we need all the machinery, all the capital and all the power that have been collected about this most difficult phase of the food question, and to take the next step we need counsel and help, not prosecution and distrust. The writer holds no brief for the packers, but the farmer will oppose, and rightly, the reckless demolishing of the only machinery that exists for handling the most costly, the most perishable and the most hazardous of all forms of human food.

The packer says there is no help for this state of things, in which the packer is wrong, because no other business of the magnitude and hazard of the cattle business is conducted totally without a contract basis.

Here is food for inquiry and fruitful opportunity for amending and improving our business—not for tearing down what is greatly needed or for putting successful men in jail for what they have not done. We need them to run the business of feeding the world.

Another example of the hasty reaction of the public to a serious situation is in the case of milk, the production and price of which constituted one of the first and most serious of all the food problems to arise under war conditions.

The scarcity of labor due to the draft and the call from munition factories and other industrial plants combined with the reduction in feed due to the shortage of wheat and the loss of a good portion of the corn crop by frost, all made the maintenance of an adequate supply of city milk in 1917 extremely difficult and the increased cost both of feed and labor made the old prices impossible if any milk at all was to be furnished.

Misleading Surface Facts

Unfortunately the hard conditions became acute in the autumn, the most unfortunate time of the year. In some cities the price was advanced and the supply was fairly well maintained. In others it was not advanced, leaving the farmers only one resource this side of bankruptcy—namely, to sell off part of the herd and cut down the feed upon the rest, both of which reduced the supply and automatically raised the price.

In cities of this kind, notably in Chicago, the public closed its eyes to the fundamental facts of which they had been well advised and proceeded upon merely surface indications, which were well calculated to deceive the very elect and about which intelligent laymen should have had judgment enough to accept the statements and advice of those who really were familiar with basic conditions.

For example, the public reasoned, "What is the need of raising the price at the very beginning of winter when there is already more milk coming into the markets than the city can consume?" That was a surface fact. The real underlying fact was, and always is, that, in order to supply a city with milk the year round, the cows must be fresh in the fall, and that means a surplus in November and December if enough is to be had later on, because the best cows with the best of care gradually diminish in their flow as the period of lactation advances.

Again the public reasoned, "Butter is made from milk and therefore there should be no discrepancy between the price of milk for city consumption and the open-market value of the butter that could be made from that milk"—a perfectly taking argument, but founded upon surface facts only. The real fact is that butter is mostly made from milk produced in the summer months from cows running on pasture lands too rough for farming and from the small surplus on the thousands of farms where enough cows are kept to supply the family needs even in the short season.

Butter, therefore, is really a kind of by-product of general farming and of rough lands or, at least, enough of it is of this character to fix prices, while city milk is produced mainly in the winter season as a primary business on valuable lands not far from the city with inspected herds that are fed on costly feeds.

Instead of facing these facts in advance and raising the price from twelve to thirteen cents a quart, the Chicago public indicted the officers of the Milk Producers, Association for attending a meeting of the producers where prices were discussed, and they are now on trial under the criminal law. Incidentally, Chicago soon paid fifteen cents a quart, due to scarcity, all of which might have been prevented by a decent attention to fundamental and readily ascertainable facts.

After-the-War Fighting

The courts may convict half a dozen officers upon a technicality, but they cannot put 13,000 dairymen in jail, or if they do, how will Chicago babies be fed? The principle of collective bargaining must be recognized, but it must not be abused. Where the farmers were wrong was not in meeting to discuss the situation in which they found themselves, but in declaring a milk strike. If any milk was wasted at that time, as has been averred, that offense is and always should be punishable by law.

The strike at best is a horrid weapon, to be used only as a last resort. It is being vastly overworked at this particular time and if continued will undermine society. The farmers were wise in not continuing the strike, but in accepting the conditions and adjusting themselves accordingly.

The weakness of the steel strike lies in the fact that, as far as we can see, irresponsible labor seeks to obtain control of one of the greatest of the basic industries, not to make life and work tolerable, but for the power it will gain. To get this power it holds up society by freezing and starving it into submission.

This strikes at the very heart of civilization and the very foundations of society and good government. Thinking farmers will not imitate these methods, for they do not seek power, but our industrial salvation must be worked out together and along lines that will harmonize with and not destroy the foundations of good government and a stable society, the legacies of our forefathers.

What has become of the splendid spirit of patriotism with which we fought the war? Where is the unity of action with which we met the Hun at Chateau Thierry, in the jungles of the Argonne, in the whirlwinds of the air and upon the highways of the sea? Where is the vision of better times to come and of better things to be? Has it descended into a series of local contests between political, industrial and social factions, making faces at each other and pulling each other's hair under the foolish pretext of setting the world in order?

Clearly, this after-the-war fighting is not of the soldier's making. He has had enough, and makes few complaints either of his broken business, of the deprivations of his family or of his own hardships and dangers. The unrest of the world is mainly the unrest of the civilian who is either unable or unwilling to adjust himself to changed and changing conditions, and by reason of this the Great War has broken up into bits of controversy enough to threaten the very stability of institutions the war was fought to preserve. Let us declare peace with ourselves and correct our derangements by a careful study of all that is involved. We have had enough of violence and overturning of long established institutions to last us for a hundred years. Let us now be sensible.

Meteorological Report, Longmont

FOR OCTOBER, 1918 AND 1919

TEMPERATURES:	1919	1918
Mean Maximum	60.5°	*
Mean Minimum	27.6°	44.50°
Monthly Mean	44.1°	*
Departure from Normal	−3.6°	*
Maximum	80.0° on 1st	*
Minimum	12.0° on 27th	17.00° on 26th
PRECIPITATION IN INCHES:		
To Date	8.90	17.25
For Month	0.85	0.42
Greatest in 24 hours	0.63 on 3rd	0.24 on 18th
Departure from normal for mo.	−0.45	--0.90
Departure from normal since Jan.1	−5.06	
NUMBER OF DAYS:		
Clear	19	10
Partly Cloudy	4	17
Cloudy	8	4

*Omitted on account of faulty maximum thermometer.

Economic Relation of the Horse to Our National Life

Wayne Dinsmore
Secretary Percheron Society of America

 WELL-KNOWN agricultural engineer asserts that the available power on farms is greater by far than all which exists in manufacturing industries. Most of this power is in horses. From this standpoint equines are power units, here to work. We use horses for scarcely any other purpose, although there is a salvage in the form of hides and fertilizer that amounts to a good deal in the aggregate.

As power units, horses come into competition with every other source of power. If they are to continue to hold a place in the world, it must be because they have certain advantages over any other type of motive power. As horsemen, it is our duty to determine these advantages and to make known sound economic reasons for continuing the production and use of horses for power purposes.

The Relation of the Horse to Human Labor

Horse power (in the form of horses) is produced with less expenditure of human labor than in any other type. On the prairies of the west colts are conceived, foaled and matured without the intervention of any human labor whatsoever. Even on corn belt farms, human labor enters but slightly into the production of horses, for where intelligent management prevails the growing colts run on pasture from May 1st till December 1st, and are out during the day for the remaining months of the year. The additional feed required per colt during the five winter months—December 1st to May 1st—will be amply supplied by the crops grown on one acre together with winter pasture. One acre of pasture—half in sweet clover and half in bluegrass—will grow a colt well from May 20th till November 1st, and another half acre of bluegrass pasture is reserved through the season to turn on about November 1st, being used, with the roughage and grain from the one acre devoted to crop, to carry the colt through the winter. It follows therefore that it requires, to grow a colt from foaling till 36 months of age, the use of 2½ acres for 3 years, or 7½ acres in all.

The horse and man labor required to grow the grain and forage crops on 3 acres of the above amounts to 8 days of horse labor and 4½ days of man labor and to this we must add 12 days more man

617

labor—4 days per year—spent in fertilizing and seeding pastures and in feeding and caring for the colt during the winter.

These figures are based on crops which are being produced by good live stock farmers, and pertain to land producing per acre 50 bushels oats, and 1 ton straw, 60 bushels corn and 1½ tons stover, and 4 tons of alfalfa hay, part of which is exchanged for bran at the rate of 2 pounds alfalfa for 1 pound bran: and these yields are being secured by thousands of good live stock farmers in the middle west. The days' labor required are figured from actual time records of farm operations worked out by the University of Illinois.

The total cost of a good grade draft colt at 3 years of age in acreage and labor therefore amounts to the use of 2½ acres of land 3 seasons, or 7½ acres in all, at such rent as prevails in such a community, plus 16½ days of man labor and 8 days of horse labor. To this cost we must add the service fee of the sire, $20.00 and still further, 20 per cent of the foregoing cost, which we add to allow for the individual colt's share of fence and shelter, seed and machinery, possible loss of colts after some expenditure has been incurred thereon, veterinary expense, taxes and other incidentals. This gives us the total cost of the colt at 36 months of age; but against this we must credit the fertilizer produced, which will be discussed later.

Limited labor is required also in the maintenance of a horse at hard work on the farm. Under conditions similar to those cited for growing draft colts 2 acres in crop—corn, oats and alfalfa—together with one acre in pasture, will carry a hard-worked draft horse a year on the farm and provide an ample allowance for a horse weighing over 1,600 pounds in working flesh. Reducing this to terms of men and horse labor, we find that it will require 5 1-3 days of horse labor and 3 days of man labor to produce the 2 acres of crop. The labor required to care for the horse while at work may be ignored, for it is covered by the general labor charge, and is less in any event, than the labor required with any other type of motive power, and much less expensive, for specialized, highly paid labor is required to operate other types of motive power satisfactorily, while the horse will operate generally with good results even though handled by boys or indifferent employees.

Labor required in caring for the horse when idle, but not out on pasture, must be provided for, however, and this amounts to about 8 days man labor per year: so that the cost of maintaining a horse at hard farm work consists of the use of 3 acres of land 1 year (at such rent as prevails in communities where such productive land is found), 5 1-3 days of horse labor and 11 days of man labor. To this, in reckoning costs, we must add 10 per cent of the preceding cost to provide for extra horses carried during the busy season, $20.00 for insurance, $10.00 for shoeing, and $20.00 for the annual share each horse must contribute to shelter, harness, taxes, depreciation and other incidentals; though it must be stated that under good farm management horses are sold before much, if any, depreciation occurs. Against this total cost, the fertilizer produced must be given credit at its actual worth.

Reduce Labor Expenditure

Our limiting factor at present in agriculture and industry is labor. We have already seen that the production of a colt to working age—36 months—or the maintenance of a hard worked horse for one year does not require much human labor.

The production of other types of motive power calculated to do work which the horse does in agriculture or commerce requires a much greater labor expenditure. Prof. W. F. Handschin, chief of the Department of Farm Organization and Management, at the University of Illinois, declares, after 7 years cost studies on Illinois farms, that mechanical power, even on larger farms, has not displaced more than 20 per cent of the horses used. If we were to assume, however, that it might displace one-third of the horses, it would displace but 4 horses where 12 were used. Mr. Parsonage, a well-known engineer, states that testimony taken before the War Priorities Board in 1918, indicated that it required at least 150 days labor to produce the gasoline tractor, which is, at present, the most widely advertised competitor of horse power on the farm. We have already noted, however, that to produce and rear 4 horses to working age, requires, in man labor, but 66 days; and it must also be remembered that farm labor is much less costly than the highly specialized labor required to produce types of motive power other than horses. The use of horses for motive power purposes on the farm and in transportation therefore reduces materially the burden on human labor and the cost thereof. It may be argued that we should expend this additional human labor rather than to give up the use of the land, but to this the answer is that it is lack of labor, not lack of land, that is the most pressing problem now; and that this will be true for so many decades yet to come that such a contention is of academic interest.

Avoids Labor Concentration

The production of power in horses keeps labor widely scattered, close to food sources, and the farmer free from possible arbitrary action on the part of employer or employees which might suddenly, even arbitrarily, double or treble the cost of power, or cut off that power entirely. Numerous instances in point could be cited but two will probably suffice. Viall Bros. of Chicago, who assemble some motor trucks on order, recently stated that in 1914 it took but 260 hours to assemble a motor truck with labor costing less than 40 cents an hour; but that in 1919 it required, to assemble a truck of identical character, in the same shop and under the same conditions, no less than 500 hours labor, costing 85 cents per hour. This quadruples cost. Another instance was noted in the testimony given by the packers, before Judge Alschuler, in the wage controversy, where it was shown that wages had increased, per hours, 132 per cent, but that the production per hour had fallen off 25 per cent; and this tendency to try to get a living without rendering a fair return in labor creeps in more and more where labor is concentrated in large cities or factory towns. The use of horses wherever they can be as economically utilized, in agriculture or commerce, as other types of motive power, has therefore, the advantage of reducing the drain on human labor, and also tends to keep it in the country close to sources of food supply, and out of concentrated districts where unfavorable living conditions inhibit family health and well being.

The Horse An Efficient Power Plant

The horse is—next to man himself—the most efficient power unit in existence, delivering more effective motive energy in proportion to energy consumed than any other type of motive power unit, when the work done as a self reproducing, self repairing organism, is taken into account. Millions of horses have worked from the time they were 3 till they were 12 years old, without the expenditure of a dollar for repairs; and this factor of long life must be taken into account in reckoning the efficiency of a power unit, for one which wears longest and with least expense for repairs has an appreciable advantage. From the economic standpoint, therefore, the horse requires a minimum of human labor in his production, and has the merit of long life and low repair cost—factors important to low cost of production in any enterprise in which power in the form of horses may be used.

Wherever power is needed to move loads over fields or roads, emergencies arise where the power required to move the load becomes three or four times normal. Horses excel in such emergencies for they can, in a pinch, exert a tractive pull equal to more than three-fifths of their live weight, or can, for a short time, pull an overload of 300 per cent to 400 per cent. In this the horse is unequalled, for no other type of motive power can handle more than a 100 per cent overload. This capacity to sustain an overload is of incalculable value in field work, especially in the spring season, when fields may be in perfect condition for work, save for occasional irregularly distributed soft spots. Horses go through these with ease, because of their reserve power, and this gives a reliability possessed by no other power unit used in field work. In city work, also, particularly on cobblestone paving, a pair of big drafters can handle an 8-ton load on a 2-ton truck solely because of the overload capacity they possess, which enables them to start the load, 10 tons in all, which, once started, can be drawn without difficulty. This ability to exert 3 or 4 times the pull usually required is therefore a distinct economic advantage.

The great flexibility of power in horses is especially valuable on the farm. One eight-horse team on a double disc with a harrow behind, may later be broken into two four-horse teams for seeding or into one pair for planting and a four for harrowing, and an extra pair for general work; or a little later into four separate teams for cultivating. No other source of power in actual use on the farm has this flexibility; and the same applies to hauling for, when six-horse teams are needed on heavy roads, they can be used readily, but can be broken into three teams and put on three separate jobs when necessity requires.

Contribution of Horse to Fertility Maintenance

No consideration of the financial cost of rearing colts, or maintaining work horses can be complete without due allowance for the value of the fertilizing constituents present in the excreta. In the case of growing colts, little is lost through leaching or poor handling for three-fourths of their time—or nearly that—is spent on pasture and the fertilizing constituents go back with very little waste. Work horses spend half their time or more in the fields at work, out on pasture nights when working, or on pasture during the idle days; so that here again losses are limited to a portion of the droppings: and this is of considerable importance.

The fertilizing constituents contained in the annual excreta from a growing colt will, at present prices, cost $66.00 if purchased in commercial fertilizers: while a mature draft horse (1,600 pounds in weight) will produce the equivalent of $88.00 worth. These are theoretical figures, however, and the farmer is interested in knowing what crop gain he can count on. The draft colt we have considered, will average 10 tons excreta per year; the draft horse 13 tons. The value of this in increasing crop yields has been indicated by the Ohio Experiment Station, where Director Thorne reports that where 8 tons per acre of barnyard fertilizer was applied every three years, the gain in crops over a nine-year test period, demonstrated that each ton of such fertilizer gave an annual gain for nine years of 2.4225 bushels corn, 134.5 pounds stover, 1,24625 bushels wheat, 128 pounds wheat straw, and 163.625 pounds clover hay. In other words each ton of barnyard fertilizer, so used, produced a gain in crop of the amounts of grain and forage named: and on present-day prices this would make each ton worth $8.20 in increased crops. On this basis the gain in crop would be worth $82.00 in the case of the colt, and $106.00 in the case of the work horse.

These figures, however, are not comparable to our needs, for Thorne's experiments were carried out on land of low fertility, which produced, untreated, but 37 bushels of corn per acre: and no such gain could be expected on land as productive as that we have considered in figuring our crop yields. Some gain would unquestionably occur, however, especially if the fertilizer produced were supplemented with a moderate amount of limestone and raw rock phosphate. One Illinois farm so handled, has increased corn yields from 35 to 80 bushels per acre within 15 years, and the expenditures for phosphate and lime have been trifling in amount: and another Illinois farm where horses have been reared in numbers for 30 years, and where barnyard fertilizer alone has been used, produced in 1917, 115 bushels of oats to the acre. These instances indicate the possibility of increases in yield through the use of the fertilizing constituents found in the annual excreta from horses kept on the farm. How much gain in crop will result, or how much the material is worth per ton, depends entirely upon local conditions. It is safe to say, however, that with the alfalfa grown and fertilizer produced, the land devoted to horse production will gain slowly but surely in production, and some surplus fertilizer be left over for use on less productive portions of the farm, where its effect in increase of crops will be still more marked. Experienced farmers and farm management experts therefore consider such barnyard fertilizer from horses worth, at present prices for farm products, at least $4.00 per ton, or $40.00 annually for the growing colt, and $52.00 for the horse at work. Men who ignore the value of such barnyard fertilizer will soon find their crop yields failing, and the purchase of commercial fertilizers will make the prices named look moderate: for there is a humus building, bacteria stimulating effect in barnyard fertilizer that is found in no commercial fertilizer, and this is a factor which makes the price just named reasonable: but the rule by which each man must determine the value of a ton of barnyard fertilizer from horses, is—"What will it produce in increased crops, or in yields maintained, compared with depreciation in yields where not used?" By this test each man must determine its worth, remembering that a growing colt produces 10 tons, and a work horse 13 tons, annually.

Horse Labor Actually Low in Cost

The total cost of producing and rearing a draft colt to 36 months of age depends first of all on the rent per acre charged for land. Ground productive enough to yield such crops rents in Illinois for about $10.00 per acre: labor including board, costs about $3.00 per day, and horse labor not over $1.50 per day. On such costs, after taking into account all factors previously discussed as entering into the cost of a draft colt, we find the total to be $187.00 against which we have a credit of 30 tons of fertilizer, which even if valued at but $3.00 per ton, cuts the cost of a draft colt, at 3 years of age, to $97.00.

The work horse if figured on the same rate for rent and labor, plus all other factors previously discussed as entering into the annual total, will cost $128.00, less 13 tons of fertilizer at $3.00, making the net annual cost of a good draft horse on the farm only $89.00; and the allowances made for feeding the colt, and the work horse, are most liberal.

Horses Utilize by Products

It must not be forgotten that while we have charged every item of cost on the basis of a full utilization of the land and all products thereof, there is, on all cornbelt farms, a good deal of forage which it is not profitable to attempt to market save through live stock, such as stubble pasture, corn stalk fields, and grass near fence rows and field edges. This feed is utilized by idle horses, and while it could be utilized by beef cattle or sheep, it is safe to say that but few men will have enough cattle or sheep to completely utilize such by product feeds; so that counting such by products so utilized on good farms by horses, the cost will be less than we have estimated.

A low power cost is favorable to efficient production and the general use of horses in all possible agricultural work will not only help to maintain fertilizer, but will also make possible production of foodstuffs at the minimum cost, provided the horses are used in large units as they can be wherever the owner has a sincere desire to obtain maximum results from his man and horse labor.

Providing Surplus Power for Peak Load

On every farm the peak load in power requirements comes in the summer; and in the corn belt, it occurs in June in corn cultivation. Providing surplus power for this, on an economical basis is a problem in good management. On 240-acre farms where mixed corn belt farming prevails 5 teams of mares, aged 3, 4, 5, 6 and 7, respectively, with one team of 8-year-old geldings, will be ample. Four of the mares will, on the average, have living colts at side: the other 6 mares, being dry, can do as much per pair as a team of geldings and they, with the 3-year-old geldings, provide 4 teams that can be in the harness every day. The 4 mares with foals will do the work of one pair all the time, and in a pinch, both pairs can be put into harness for a few days without any particular injury to the colts: but as a matter of good practice horsemen like to give mares with foals half time off: and this can be done most of the time under the plan outlined above. Each fall the mares past 7 years of age should be sold unless some younger mare has proved persistently barren, in which case she should go instead: and the young geldings, now 3½ years of age, should also be disposed of, so that in this way, the peak load requirements are taken care of, without carrying a surplus of horse power throughout the year: and this is all important, for while it is desirable to have ample surplus through the busy season, the extra power—in such a case two pairs—should be disposed of as soon afterward as possible, even though the price may not be quite as high as is desired: and in making such sales, the actual cost of production should not be overlooked, for it is under most conditions considerably less than is usually estimated.

Effect Upon National Life

Any great shift in the use of horses as power units must have far reaching, incalculable effects upon our national life. More human labor must be used in iron and coal mines, on vessels and railroads, in smelters and steel mills, and in the factories where other type of motive power—be they gas, steam or electric—are finally fabricated. This draws more heavily upon our existing supply of human labor, calls more men from farms to cities, mines and factories, drives labor higher and higher in price, and curtails the production of other things, useful to the world, which might have been made with the labor devoted to manufacturing motive power units designed to do the work the horse can do, and does do, more efficiently and more economically than such horse substitutes.

Last, but not least, is its effect upon our citizenship. The production of power units, in the shape of horses, has its influence on family life, and in liking for the farm. Boys are educated to care for the colts, and to look for improvement in each generation: and in that training they are taught, step by step, the principles of breeding—that like produces like—that faulty parentage will show in the progeny to the third and fourth generation—and that development of the possibilities inherent in an individual requires environment making this possible. The use of horses as power units on the farm interests farmer and family in the production of other good live stock: and stock production inculcates in children a liking for the land, thought as to how its fertility may be increased, and a desire to breed and rear animals that are more efficient in work or in meat production than any one else has produced. No such training comes about where other types of motive power are used, for, on the contrary, the training required along mechanical lines leads a great many youngsters into factory work, where, attracted by the lure of apparently high wages, they are lost to the farm for all time. In the end, when rents and high living costs are paid, they are usually worth but a fraction of the wealth possessed by their brothers who remained on the farm.

All industries are today so inextricably woven together in our national life that anything which affects one basic industry affects all, and no policy can long endure which is not based on sound economic principles. Attempts to displace horses in agriculture and commerce increase the burden on human labor, and on costs of production. Continuance of the use of horses as motive power units to the utmost in agriculture and transportation will release hundreds of thousands of workers to other lines of work, or permit them to return to agricultural pursuits, and will lessen the drain on farm labor which has for 3 years past crippled efficient farm operation. It will keep more families in the country close to the source of supplies, and will endure because it is sound in principle and in practice. More homes, more owner operated farms, and a better social life will ultimately result.

Is it desirable for any nation to have 5-6 or 7-8 of its population centered in mines, factories, and cities, leaving but a small proportion on the land? Are boys and girls healthier and happier in the city? Are they better educated in all that pertains to life? Are they as thoughtful, steady and sane as the boys and girls on farms? Will they average as high in all that goes to make good citizenship? It is not our task to answer these questions. Statesmen of our nation have answered them for us. They have pointed out, over and over again, that the leaders in manufacturing, transportation, commerce and war are usually farm raised boys. They have called our attention to the fact that as soon as a city man acquires a fair competence, he wishes to buy a farm where his children can enjoy that which farm children have as their heritage—open air, the feel of the ground beneath their feet, and the companionship of living, growing plants and animals. From the days of Washington our great statesmen have recognized that farm life is favorable to clear thinking and right living: and today this is more than ever true, for in the clean environment of country life, in modern homes, and with special attention to sound education, farm children develop into clear cut, clean minded youths, who are, as our statesmen oft remind us, the bulwark of our nation—well educated, clear thinking, sane Americans.

Get in Line with Pure-Breds

E. R. McIntyre

NE of the objections which the better live stock committees are meeting in their campaign in various counites to increase the use of pure-bred, production sires, is the fixed idea of some farmers that pure-bred bulls are just another added expense—a luxury which they cannot afford and which is really beyond their means. Perhaps there are a few men here and there who cannot put their hands on ready cash right away with which to invest in good bulls, yet even this difficulty can be overcome by the "trade-even" or "long-time-payment" plan. Banks are usually anxious to be identified with the better bull movement and are doing a great piece of promotion work in the extension of credit facilities to worthy

farmers. There is still another way to arrive at the end desired, and
that is to form community bull clubs, like this one described by the
Missouri Agricultural College:

"About a year ago Webster County dairymen found it had been
costing them $19.75 a year each for the service of a $75 bull, while
each man had a "scrub" bull around his place. After organizing a
co-operative bull association it cost them only $5.50 a year each for
the use of a $260 bull, and in addition they have only seven bulls to
maintain while before organizing they had 18.

"The Webster County dairymen had been getting along as best
they could with 'scrub' bulls, each feeling that he could not afford
to pay the price of a high-grade bull. They came to the point, however, where they realized they could not improve their herds until
they stocked up some well-bred bulls. Then, to answer their need,
38 Jersey breeders met and formed a Jersey bull asociation. They
divided themselves into seven well-arranged communities, or 'blocks,'
and purchased a bull for each 'block.' Each man was assessed $6.00
for every cow he owned and seven Jersey bulls, whose dams for four
generations back averaged 622 pounds of butter, were purchased.
Every two years the bulls are alternated in numerical order. So that
with the original cost price a man has the use of such a bull for as
long a time as these bulls are serviceable.

"But the economy of bull service is only a minor point in comparison with other advantages these men are deriving. Through the
use of these bulls they are building up their herds to a much more
profitable standard. More than this, most of them are working
gradually into the pure-bred business and are building toward a
strong Jersey center. In this way, as a community, they can advertise and attract buyers that no individual could ever hope to do."

Webster County has made a move that any Colorado community
can do equally well. It is a practical business proposition that works,
and it works to the advantage of all concerned. Next to owning a
little pure-bred calf that may develop into a fine sire in a short
time, our farmers who are raising cattle indiscriminately could hardly
arrive at the end desired more economically than with the bull club
plan. This is especially true in northern Colorado regions in communities where the dairy business is just getting safely started.

The Hired Man's Side of the Story

EVERYWHERE one meets farmers who tell you how hard it is to get and keep good, intelligent, competent farm hands. The papers are full of articles on the farm labor problem; the windows of banks, postoffices, and stores display placards urging young men to go back to the farm.

What is the trouble? Why can't farmers get and keep the class of farm help they deserve? We hear so much about it these days that I am going to write some observations after several years of farm work.

When I first started working as a farm hand the average wage was from $30 to $35 a month and board for single men and the same amount with house, garden, milk, and fuel for a married man. At that time one could buy a good substantial pair of work shoes for $3 to $4 and an excellent pair of overalls for 75 to 90 cents. At present the average wage has gone up to $50 or $55 with a much poorer grade of shoes selling at $5.00 a pair and an equally poor grade of overalls selling at $1.75 a pair. With married men the wages have gone up even less, while the cost of living has gone up so that a married man getting the average wage—plus house, garden, milk, and fuel—can just barely make a living.

Right here, it seems to me, Mr. Farmer makes a mistake. All he expects a man to do is to make the barest kind of a living. How can he expect a man to give him a good, hard day's work when by doing so he is simply making both ends meet without being able to lay aside a little something for the rainy day or for a modest start for himself?

One string which pulls many, both married and single, away from the farm is the offer of larger wages and shorter, more regular hours. If it were only the lure of larger wages a man would be foolish to heed the call, for rent, fuel, and the other expenses incident to city life soon eat up the difference between the city and the farm wage. It is, however, the shorter and more regular hours that make many leave the farm. Conditions are such that it is impossible to shorten the farm day to eight or even ten hours, and I do not think that any reasonable, thinking farm hand expects anything of the kind, but when it comes to expecting a man to work 15 or 16 hours a day, and for a bare living, I do think it is asking a little too much.

I am very much like the Iirishman, who, when asked what he thought of a new local law closing the saloons at 10 p. m. said: "It is all right. The man who does not get drunk by 10 o'clock isn't trying." I believe that the farm hand who can't give his employer a good day's work, including chores, from 5 a. m. to 6 p. m. isn't trying either.

Farmers will tell you they would rather hire a married man whenever possible, and when asked why, will give the time-worn excuse that married men are steadier and less apt to move. True enough! The average married farm hand isn't able to save enough on which to move.

Some farmers will advertise for a married man, $50 or $60 per month, board and room for both—"wife to help with the housework." Help! Said help consists of the hardest kind of housework from early in the morning until late in the evening, and for which she receives just her board and lodging, not a cent of wages. The same employer, if unable to get a man and wife, will hire a single man for $50 or $60 a month and board and then hire a woman or girl and pay her $20 or $30 a month and board to help with the housework. However, when able to get a man and wife he forgets about the girl's wages and has the man's wife do the same work for the magnificent sum of nothing and the privilege of being near her husband. Do you think it is exactly fair?

Then there is the farmer who hires a married man for $50 or $55 a month plus house, garden, milk, and fuel. To the uninitiated, the "H., G. M., and F." sound like a lot. Let us look at them separately.

The house is uually an old affair of bygone days, sadly in need of repair and often very dirty. The plaster is broken in spots, the wall paper torn and spotted, the pantry a mass of mice holes, and frequently the house is alive with bedbugs, cockroaches and mice, and back of all stands the tumbled down, patched up, unspeakable outhouse.

As to the garden, any farm which is large enough to require the services of a hired man ought to be able to spare an acre of ground in which the tenant could raise, on his own time, a little more than some green vegetables for summer use.

The usual amount of milk furnished seems to be one quart per day, which seems to me to be an inadequate amount for the average family, and it has been my experience that in the majority of cases even this is given in such a begrudging spirit that one is made to feel as though the farmer were bestowing a great favor.

The fuel item is one which seems to cause a considerable number of farmers and hired men to part company. Many farmers say they furnish fuel, but when you start to work for them you find that the firewood is still standing in the wood lot and you are expected to get said fuel on your own time. I don't think that many hired men object to splitting some wood occasionally on their own time, but when it is necessary to give up several evenings or the whole of a Sunday to getting the wood out of the wood lot and then having to spend several evenings in bucking it up to stove lengths and splitting it, that is asking a great deal. Even if the hired man does get up the wood on his own time many farmers will everlastingly complain about the amount of wood the man's wife burns.

When a farmer gets a new bull, cow, hog, or flock of chickens he invariably cleans out and disinfects a place for them, making their new surroundings as clean and comfortable as their several tastes may require. Why not accord the new hired man and his family the same treatment?

One thing which many farmers do which is not conducive to good work is to come to where you are working and apropos of nothing, tell you what a wonder the last hired man was, how much work he used to do and how well he used to do it. It always seems foolish for a man to do that for it shows a lack of business astuteness on his part. If the last man was such a wonder, why didn't Mr. Farmer realize the fact in time, pay him what he wanted, and keep him?

I have observed the above conditions in my ten years of farm work. In that time I have worked as a farm hand, foreman, and manager. I have worked for men of wealth and education as well as men who have been born and raised on the farm and have attained only a modicum of success as success is rated in this world. These are not only my own experiences, but are the experiences of many others with whom I have worked and talked. Men, I mean, who are rated as good farm laborers by their employers and their neighbors.

Another type of farmer which is beyond my ken is the one who advertises for a manager, said manager to have a broad knowledge of farming, stock raising, be a good calf raiser, etc., etc., and have a college education. Wife to cook for so and so many men. What is the use of a man acquiring that knowledge and education if he has to fall back on his wife's ability to cook in order to get and hold said position as manager? It seems to me that a man who can hold a position like the above ought to be worth money enough on his own merits to keep his wife from any additional kitchen drudgery.—Hoard's Dairyman.

The Utility of Beet Silage

Last year I siloed some of the beet tops on my farm as an experiment. I gathered about 125 tons of beet tops from twenty-five to thirty acres of beets. I did not take pains to gather the tops up clean and had considerable pasturage left on the field. The siloing was finished October 25th and the feeding began about December 1st. I find that this silage fully comes up in feed value to all estimates.

We fed 800 head of sheep, about half ewes and half lambs, on a ration of 3 pounds of beet top silage a day for forty days and they made a gain of 4 to 5 pounds a head more than the other sheep of a like quality fed entirely on hay and beet pulp.

This three pounds of silage replaced about one-third of the hay and pulp otherwise necessary and the sheep gained flesh much faster than when fed on other feed we have used. When you figure the present price of muttons you will find that this extra gain is worth a dollar or more a head. Another thing that should be considered is that it is a very convenient feed to handle in cold weather, it is always warm and the sheep take to it very readily. The cost of siloing the tops did not exceed $1.50 a ton, including the construction of the silo, which is 100 feet long, 5 feet deep and 14 feet wide. The cost of $1.50 a ton includes the covering up of the silo.

The silage came out without any waste whatever. We sprinkled a little straw in the bottom of the silo to keep the tops from coming in direct contact with the earth, and to about every foot of silage we put in a sprinkling of two to three inches of straw; on the top of the silo we spread a little straw and then from 6 to 8 inches of short manure, which seemed to fill the bill entirely. So far as we could judge we took out about 80 per cent of the tonnage put in. I figure that this silage has a value of fully $10.00 a ton as compared with hay at $25.00 a ton. It therefore can readily be seen that getting five or six tons of tops to the acre when siloed is more valuable, acre for acre, than alfalfa, figuring alfalfa on a basis of 3 to 3½ tons to the acre, in a season. The cost of labor in putting away an acre of beet tops is much less than raising and harvesting an acre of hay.

Sugar beet growers should not lose sight of the value of the beet top silage. It is an asset that adds very materially to their profits and provides an opportunity to feed two or three carloads of cattle during the winter from which a handsome profit can be made; besides which, the manure from the cattle yards is available for fertilizing their lands. The great trouble with so many of our small inter-mountain farmers has been that when their fall work is over they have nothing further to do until planting season opens up in the spring. To be successful farmers, they should have some profitable winter work, and there is no better way of making a few hundred dollars during the winter than feeding fifty to seventy-five head of steers, as it is a well-known fact that a steer is worth from $40.00 to $50.00 more on the first day of March than it is in November.—Utah Farmer.

Eight Tons of Manure per Acre Every Four Years

TRIALS made on the various experiment fields throughout the state by the University of Missouri College of Agriculture have shown that an average application of eight tons of manure to the acre once in four years has increased the yield of corn 10.5 bushels, oats 5.17 bushels, wheat 5.24 bushels, and clover hay 937 pounds. At prices which prevailed the first of the year, this increase would be worth $4.83, and at pre-war prices $2.34, for each ton of manure applied. It will cost the farmer not more than one dollar a ton to collect the manure and haul it to the field. This would leave a net profit of $3.83 at present prices, or $1.34 at pre-war prices, for each ton of manure applied. The full value of the manure is usually not obtained during the first four years, for it leaves the soil in better condition and its effect upon later crops is often quite significant. This becomes more noticeable after the first two or three applications, since a liberal application of manure every four years will result in permanent improvement to the land.

Return Two Tons a Month

By very careful handling of manure, a live stock farmer, on average soil, should be able to return annually, about two tons of manure per acre to his cultivated fields. It is not easy to save this amount except by very careful methods and persistent efforts. It is necessary that all straw and other suitable materials be worked through the barns as bedding. This not only adds to the comfort of the animals, but serves as an absorbent for the liquid manure. If straw piles are sold, burned or left to rot, it is, of course, impossible to return this amount of manure.

Keep the manure together; don't let it get scattered about the barn or lots. Hogs and chickens may waste much of it. Where possible, haul it to the field as soon as produced. If this is not possible, store it in shallow concrete lined pits to prevent leaching. Use plenty of bedding in the stable or on the feeding floor to absorb the liquid. Straw, old hay, grass and leaves may be used for this purpose.

Manure Should Go in Pit

About 35 per cent of the nitrogen and 55 per cent of the potassium is to be found in the liquid material. There is, however, practically no phosphorus in this part. The plant foods in the liquid are all in soluble form and are very easily lost through leaching. Furthermore, the nitrogen in this liquid portion readily passes off as ammonia when the manure is allowed to ferment, hence the necessity for preventing fermentation as far as possible.

631

The total solid and liquid manure produced in a year by a well-fed, mature horse is about eight tons, with a plant-food value of more than $30. In the case of a well-fed steer weighing from 1,000 to 1,200 pounds the production is nine to eleven tons, with a slightly greater total value than the manure from a horse.

Three to five months' exposure to the weather in an open lot may cause manure to lose approximately one-third of its plant food. If manure must be exposed to the weather it should be in a pit with a water-tight bottom.—Record Stockman.

Colorado Farmers No Longer Need Plant Good Seed in Defective Soil

 MORE than 6,000 samples of seed have been tested free of charge at the laboratory of Colorado Agricultural College, Fort Collins, in the two years that the law providing for this work has been in effect.

That appears large but it really is a beginning. Farmers have been slow to realize that they need not waste their time planting indifferent seed through lack of knowledge of exactly what it is. The state will test it for them and not charge a cent.

Agricultural college officials are undertaking a campaign to advertise this fact and induce Colorado growers of grains to seek their aid in avoiding defective seed. There are penalties in the law for its infraction, which every agriculturist and seed dealer should know.

Meant for Labeling Law

The pure seed law was adopted by the general assembly in 1917, and has only been in effect a little over two years. This act was primarily a labeling law. It requires that all seed sold or offered for sale in the state shall contain a label stating the quality of the seed, the percent of purity and of germination. It provides for a laboratory, which is situated at the experiment station, and where farmers and seed merchants can send their samples of seed to be tested.

Judging from the way samples to be tested are arriving at the laboratory, more and more farmers are daily beginning to realize the value of such a law and the benefit derived from using only pure seeds on their land.

Numerous other states throughout the Union have realized the value of a pure seed law and have passed such measures. There are now twenty-nine states that have a pure seed law in some form or other. The laboratory, conducted by the state of Minnesota, tested more than 20,000 samples of seed sent to them in one year.

Samples Are Limited

The seed laboratory at Fort Collins is well equipped to take care of all samples sent in to be tested. However, the number of samples sent in by any one person or firm within the same month are limited. Not more than 25 purity tests and 50 germination tests in the same month will be made for the same person.

This law also controls the importation into Colorado of unlabeled seed. It provides: "No seed in quantities of five pounds or more shall be shipped or brought into Colorado from outside the state by any person, to be used by himself for seeding purposes unless such seed shall have been tested and the containers of such seed shall have affixed thereto a plainly written tag or label giving the information and results of the tests required by the regular label."

Many Weeds in Samples

There is a constantly increasing number of weeds to be found on the farms of Colorado. In many sections of the state, these weeds have greatly reduced the yield of crops and caused a decline in the value of such infested land. The most common source, however, impure seeds raised within the state's borders have also contributed.

Having tested a sample of seed, the laboratory keeps the samples and a record of the test. They then send the result of that test to the person from whom the sample was received. Accompanying the report is a complete list of all the noxious and other weed seeds found in the sample and the percentage of them found. Thus a farmer or seed merchant knows exactly the quality of the seed he is planting or selling.

Inspectors Keep Lookout

With the idea of enforcing this law, inspectors have been appointed to call at any time upon those selling or exchanging seeds. If they find that his seeds are not labeled, the man offering them for sale or exchange is liable to prosecution and fine.

This law was adopted with the idea of taking a step forward in bettering agricultural conditions of the state. The farmer or seed merchant is under no expense whatsoever. Farmers can purchase low-grade seeds if they so choose, but they know what they are getting as they can see for themselves by the label attached what the tests show. It is hoped that within the near future the number of samples tested by the laboratory will have reached a large figure as farmers become more familiar with the law and the value of using pure seeds in planting crops.—Record Stockman.

The Story of an American Farmer

W. W. Robbins

I WILL tell you of the life of an American farmer—a pioneer who farmed in Ohio over a generation ago. There are many of his kind throughout the length and breadth of this land; men who have done the pioneering in American agriculture—hard-working, patient, persistent, prosperity-building, honest farmers.

Early in the last century a man came with his young wife and three children from eastern Kentucky to northwest Ohio—from a settled, schooled community to one where hunters, trappers and woodsmen dominated. He built his log cabin in the midst of 320 acres of mixed hardwood forest. His oldest son, the main actor in this story, was called upon, naturally, to bear the brunt of the work—clearing and planting and building. The barn was made of walnut, the new house of walnut, the furniture of walnut. Shoes, clothing and furniture were made in the home. The hours of work were long. At the age of 21 the oldest son was given a farm; he married, and the two started on their common career. They (and I use "they," not "he,") brought that farm to a state of perfection; they raised a family of seven children; they helped to build a rural community of high quality; they gained the respect of everyone, because of their true worth; and only last year, at the age of 85, after 64 years on one farm, this American farmer passed to a reward that must have been a high one. His helpmate does not leave the farm to live with her children, but prefers to spend her days among the familiar scenes and surroundings of her life. No doubt to her there has grown up an attachment for every inch of farm and house. The path to the barn that both have tread thousands of times, the latch on the gate, the place at the table, the books, and all—is it any wonder she would find it impossible to leave the farm! They had builded what is in every sense a home. Contrast such with the city flat, with the temporary homes of those who move from place to place, house to house.

What was the character of this American farmer? First of all he was substantial—absolutely substantial. He was good clear through. He believed in hard work—not that hard work brought more wealth, but more happiness and health. He attributed his good health, his old age, to the fact that he kept busy, and he worked on the farm until the very last days of his life.

Conservative—yet progressive; systematic in the highest degree; orderly in every respect. In his library can be found the standard literature of the world and in abundance; on his shelves are found the best periodicals and farm papers. He felt that he owed it to himself, his family, and the community not only to "keep up," but to place the best before his growing boys and girls—the best of written works. His children, seven, received all the advantages of school and college. The four boys are responsible and loyal men and citizens; the three girls mothers of families.

He always took an interest in community building. Men and affairs, here and abroad, interested him. New ideas were readily assimilated; new methods and new machinery were adopted. He practiced the good old diversified farming. His farm is ideal. He left it after 64 years in better condition than when he took it—better in appearance and yielding more per acre—as beautiful a farm and homestead as one will see anywhere in the land.

The conservation of soil fertility was an ideal with him. He found the land fertile; it was his heritage. He owed it to these who came after him to transmit that heritage undiminished. This meant crop rotation. A system once established, he adhered to it. Small grain, legumes, a cultivated crop, with livestock on the farm—these were the essentials of a permanent agriculture.

At the age of 85, our hero died. He had builded a home in the truest sense—he had helped to build a prosperous community—he stood for the right always—he had taken an active interest in civic affairs, local and national, he had passed on his inheritance untainted —left in this country seven good American citizens. Who is not willing to say that his was a successful life?

FARMERS SHOULD NOT SELL STOCK—SHOULD FEED IT

It is customary for the average farmer to sell his steers and sheep to the feeder. This is a mistake. The man who owns land, has his own feed and needs manure for his soil, should feed out at least all the stock he raises. Feeding in large numbers may be considered a gamble, but the farmer who owns land and who stays in the feeding business every year, will be more prosperous in the end than the farmer who does not feed but sells his stock to be fed by some one else. Some years there will be a loss on the feeding operation if the increased yields secured by the use of manure is not considered, but on an average over a number of years there is a satisfactory profit on the feeding operation. The man who owns land, raises his feed and especially his own stock, should by all means feed regularly every year.

The man who owns no land, has no use for manure, raises no feed and has to buy what he feeds should leave feeding alone.

Winter feeding should be more generally practiced by Colorado land owners.—R. W. Clark, Colorado Agricultural College, Fort Collins, Colorado.

Why a Long Term Lease is Preferable to the Renter

Louis Bartolett

First, a long-term lease enables the farmer to fertilize and keep the fields in the proper stage of cultivation, thereby realizing a reasonable recompense for his efforts.

No man can succeed by the one-year leasing system, with an annual move. Five moves are equivalent to a fire. No two farms require the same equipment, thus making it necessary to sacrifice or increase equipment, as the case may be. And, again, the average farmer is not content with the annual move so near.

In renting farms by the season I hardly know who does himself the greater injustice, the landlord or the tenant, as the effect is the same to both parties.

The tenant can afford to do only the necessary work to make a season's crop, thus leaving the land more impoverished for the next unfortunate. I have really known cases where tenants have spent much valuable time looking for another location, and I do not believe any farmer can be as successful in this unsettled condition.